ENERGY SCIENCE, ENGINEERING

EJECTORS AND THEIR USEFULNESS IN THE ENERGY SAVINGS

ENERGY SCIENCE, ENGINEERING AND TECHNOLOGY SERIES

Stages of Deployment of Syngas Cleaning Technologies
Filomena Pinto, Rui Neto André, I. Gulyurtlu
2010. ISBN: 978-1-61668-257-6 (Online Book)

Nuclear Fuels: Manufacturing Processes, Forms and Safety
Antoine Lesage and Jérôme Tondreau (Editors)
2010. ISBN: 978-1-60876-326-9

Thin-Film Solar Cells
Abban Sahin and Hakim Kaya (Editors)
2010. ISBN: 978-1-60741-818-4

Wind Turbines: Types, Economics and Development
Gerard Martin and Jeremie Roux (Editors)
2010. ISBN: 978-1-60741-849-8

Advanced Materials and Systems for Energy Conversion:
Fundamentals and Applications
Yong X. Gan
2010. ISBN: 978-1-60876-349-8

A Solar Car Primer
Eric F. Thacher
2010. ISBN: 978-1-60876-161-6

A Solar Car Primer
Eric F. Thacher
2010. ISBN: 978-1-61668-382-5 (Online Book)

Handbook of Sustainable Energy
W. H. Lee and V. G. Cho
2010. ISBN: 978-1-60876-263-7

Jatropha Curcas as a Premier Biofuel: Cost, Growing
and Management
Claude Ponterio and Costanza Ferra (Editors)
2010. ISBN: 978-1-60876-003-9

Ethanol Biofuel Production
Bratt P. Haas
2010: ISBN: 978-1-60876-086-2

Biodiesel Handling and Use Guide
Bryan D. O'Connery (Editor)
2010. ISBN: 978-1-60876-138-8

**Syngas Generation from Hydrocarbons
and Oxygenates with Structured Catalysts**
*Vladislav Sadykov, L. Bobrova, S. Pavlova, V. Simagina,
L. Makarshin, V. Julian,
R. H. Ross, and Claude Mirodatos*
2010. ISBN: 978-1-60876-323-8

**Corn Straw and Biomass Blends: Combustion Characteristics
and NO Formation**
Zhengqi Li (Editors)
2010. ISBN: 978-1-60876-578-2

Introduction to Power Generation Technologies
Andreas Poullikkas (
2010. ISBN: 978-1-60876-472-3

**CFD Modeling and Analysis of Different Novel Designs of Air-Breathing
Pem Fuel Cells**
Maher A.R. Sadiq Al-Baghdadi
2010. ISBN: 978-1-60876-489-1

**A Sociological Look at Biofuels: Understanding
the Past/Prospects for the Future**
Michael S. Carolan
2010: ISBN: 978-1-60876-708-3

Direct Methanol Fuel Cells
A. S. Arico, V. Baglio and V. Antonucci
2010. ISBN: 978-1-60876-865-3

Ejectors and their Usefulness in the Energy Savings
Latra Boumaraf André Lallemand and Philippe Haberschill
2010. ISBN: 978-1-61668-210-1

Sustainable Resilience of Energy Systems
Naim Hamdia Afgan
2010. ISBN: 978-1-61668-483-9

Sustainable Resilience of Energy Systems
Naim Hamdia Afgan
2010. ISBN: 978-1-61668-483-9 (Online Book

Utilisation and Development of Solar and Wind Resources
Abdeen Mustafa Omer
2010. ISBN: 978-1-61668-238-5

Coalbed Natural Gas: Energy and Environment
K.J. Reddy
2010. ISBN: 978-1-61668-036-7

Storage and Reliability of Electricity
Edward T. Glasby
2010. ISBN: 978-1-61668-248-4

Microalgae for Biofuel Production and CO2 Sequestration
Bei Wang, Christopher Lan, Noemie Courchesne, and Yangling Mu
2010. ISBN: 978-1-61668-151-7

Geothermal Energy: The Resource Under our Feet
Charles T. Malloy (Editors)
2010. ISBN: 978-1-60741-502-2

Wind Power Market and Economic Trends
Martin W. Urley (Editors)
2010. ISBN: 978-1-61668-243-9

High Temperature Electrolysis in Large-Scale Hydrogen Production
Yu Bo , Xu Jingming
2010. ISBN: 978-1-60021-121-6

High Temperature Electrolysis in Large-Scale Hydrogen Production
Yu Bo , Xu Jingming
2010. ISBN: 978-1-61668-697-0 (Online Book)

Biofuels from Fischer-Tropsch Synthesis
M. Ojeda and S. Rojas (Editors)
2010. ISBN: 978-1-61668-366-5

Biofuels from Fischer-Tropsch Synthesis
M. Ojeda and S. Rojas (Editors)
2010. ISBN: 978-1-61668-366-5 (Online Book)

**Handbook of Exergy, Hydrogen Energy
and Hydropower Research**
Gaston Pélissier and Arthur Calvet (Editors)
2009. ISBN: 978-1-60741-715-6

**Computational Techniques: The Multiphase CFD Approach to
Fluidization and Green Energy Technologies (includes CD-ROM)**
Dimitri Gidaspow and Veeraya Jiradilok
2009. ISBN: 978-1-60876-024-4

**Cool Power: Natural Ventilation Systems
in Historic Buildings**
Carla Balocco and Giuseppe Grazzini
2010. ISBN: 978-1-60876-129-6

Transient Diffusion in Nuclear Fuels Processes
Kal Renganathan Sharma
2010. ISBN: 978-1-61668-369-6

Transient Diffusion in Nuclear Fuels Processes
Kal Renganathan Sharma
2010. ISBN: 978-1-61668-369-6 (Online Book)

Coal Combustion Research
Christopher T. Grace (Editors)
2010. ISBN: 978-1-61668-423-5

Coal Combustion Research
Christopher T. Grace (Editors)
2010. ISBN: 978-1-61668-423-5 (Online Book)

Utilisation and Development of Solar and Wind Resources
Abdeen Mustafa Omer
2010. ISBN: 978-1-61668-497-6

Shale Gas Development
Katelyn M. Nash (Editors)
2010. ISBN: 978-1-61668-545-4

Shale Gas Development
Katelyn M. Nash (Editors)
2010. ISBN: 978-1-61668-545-4 (Online Book)

ENERGY SCIENCE, ENGINEERING AND TECHNOLOGY SERIES

EJECTORS AND THEIR USEFULNESS IN THE ENERGY SAVINGS

LATRA BOUMARAF
ANDRÉ LALLEMAND
AND
PHILIPPE HABERSCHILL

Nova Science Publishers, Inc.
New York

LIBRARY OF CONGRESS CATALOGING-IN-PUBLICATION DATA

Available upon request.

ISBN : 978-1-61668-210-1

Published by Nova Science Publishers, Inc. † New York

CONTENTS

Contents

PREFACE

The object of this chapter is to introduce the subject of ejector in refrigerating plants in view of energy savings.

A first part concerns the use of an ejector in a heat-operated system for the production of cold (refrigeration or air conditioning). The ejector refrigeration system using an environment friendly fluid allows combining two advantages: one related to the energy saving by the use of thermal energy wasted from industrial processes at average or low temperatures or a free energy source (solar energy) and the other related to the environment protection by reduction of CO_2 emissions in the atmosphere. The performance of this system depends strongly on that of its ejector, thus many works have been focused on this component. Particularly, it has been the subject of theoretical and experimental studies in the Thermal Center of Lyon (France) and the mechanical department of Annaba university (Algeria) during many years. In addition to a literature review, the paper provides a summary of these works, which have been previously published in several journals and presented in many international conferences. In this chapter, these studies are gathered and discussed in several sections, i.e. ejector behavior analysis in different operating modes, performance characteristics, working fluids, modeling of an ejector refrigeration system based on those of its components in design and off-design conditions.

A second part consists of an original simulation program of a transcritical CO_2 refrigeration system using an ejector as an expansion device in order to improve the COP of the basic transcritical CO_2 system by reducing the isenthalpic expansion losses. A constant-area mixing model is used to design the ejector for typical air conditioning applications. By using this ejector model, a thermodynamic cycle analysis of the transcritical CO_2 system with expansion by ejector is performed. The results highlight the benefit of using ejector as an expansion device in improving the system energetic efficiency.

Keywords: Ejector; Modeling; Simulation; COP; Refrigerant fluid; Transcritical CO_2 cycle; Expansion device

AUTHORS' CONTACT INFORMATION

Latra Boumaraf[*]
University of Annaba, Annaba, Algeria
National Institute for Applied Sciences, Villeurbanne, France

André Lallemand
National Institute for Applied Sciences, Villeurbanne, France

Philippe Haberschill
National Institute for Applied Sciences, Villeurbanne, France

---

[*] Corresponding author. Tel.: +33 472 43 79 51. Fax.: +33 472 43 88 11. *E-mail address:* l_boumaraf@yahoo.fr

NOMENCLATURE

A	constant in Eq. (II.15) or area section of ejector (m²)
A_P^*	area section of primary nozzle throat (m²)
A_S^*	area section of the secondary flow choking throat (m²)
a	constant in Eq. (II.15)
C	refrigerant mass in the ejector refrigeration system (kg)
C_P	specific heat of the gas or vapor at constant pressure (J/kg K)
C_w	specific heat of the HTF (J/kg K)
COP	coefficient of performance
COP_C	performance coefficient of Carnot cycle
D	diameter of the ejector cylindrical tube (m)
d	diameter of the primary nozzle exit section (m)
d_D	diameter of the diffuser (m)
Er	error
h	specific enthalpy (J/kg)
I	improvement of the ejector expansion transcritical CO_2 refrigeration system equal to COP/COP_b
K	constant in Eq. (I.14)
L	length (m)
L_V	latent heat of the refrigerant fluid (J/kg)
M	Mach number
M'	molecular weight of the motive fluid in Eqs. (I.14-15)
M"	molecular weight of the driven fluid in Eqs. (I.14-15)
M_F	mass of the refrigerant fluid in a heat exchanger (kg)
$\dot{M}_W$	mass flow rate of the HTF(kg/s)
M_σ	refrigerant mass in the subcooling zone (kg)
$\dot{m}_P$	mass flow rate of the primary fluid (kg/s)
$\dot{m}_S$	mass flow rate of the secondary fluid (kg/s)

Nomenclature (Continued)

P	pressure (Pa)
P_{C0}	limiting condition on back pressure of the ejector operational mode (Fig I.4.)
$\dot{Q}$	heat rate (W)
R	universal gas constant (J/kg K)
r	pressure lift ratio or compression ratio of an ejector equal to P_C/P_E
Re	Reynolds number
S	heat transfer surface (m²)
s	specific entropy (J/kg K)
T	refrigerant temperature (°C) or (K)
T_P	wall temperature in a heat exchanger (°C)
t	temperature of the HTF (°C)
U	entrainment ratio of an ejector (Eq. (I.1))
U'	entraiment ratio of an ejector with different molecular weights of the primary and secondary fluids (Eqs. (I.14-15))
V	velocity of the refrigerant fluid (m/s)
v	specific volume of the refrigerant fluid (m³/kg)
W	volume of a heat exchanger (m³)
$\dot{W}_P$	work rate of the pump (W)
x	vapor quality of the refrigerant fluid or related to a geometrical parameter of an ejector in Table I.1.
Greek symbols	
α	characteristic angle of an ejector defined in Table I.1 or coefficient of heat transfer (W/m²K)
β	characteristic angle of an ejector defined in Table I.1.
Φ	geometrical parameter of an ejector equal to $(D/d^*)^2$
φ	geometrical parameter of an ejector equal to $(d/d^*)^2$
η	isentropic efficiency
η_P	mechanical efficiency of the pump
η_{ex}	exergetic efficiency
λ	thermal conductivity (W/mK)
μ	dynamic viscosity (kg/ms)
ρ	density (kg/m³)
σ	surface tension (N/m)
Θ	filling amount of a heat exchanger
Γ	expansion ratio of an ejector equal to P_B/P_E
γ	characteristic angle of an ejector defined in Table I.1 or specific heat ratio of the refrigerant fluid
Δ	variation

Nomenclature (Continued)

ΔT	Superheat (K or °C)
ξ	driving pressure ratio of an ejector equal to P_B/P_C
Ω	geometrical parameter of an ejector defined in Table I.1.
Subscripts	
B	boiler
b	related to the basic cycle
C	condenser or related to Carnot cycle
cal	calculated value
comp	related to the compressor
cr	related to the refrigerant critical state
D	diffuser
E	evaporator
e	related to the inlet
i	incremental value
is	isentropic process
L	liquid
M	related to the motive flow
N	related to the primary nozzle
opt	optimal
P	related to the primary flow
S	related to the secondary flow or suction chamber
s	related to the outlet
V	vapor
σ	related to the subcooling zone
0	related to the stagnation state
0,1, 2,...	locations in the ejector and the operating cycle of the system
Superscript	
*	at transition mode or related to a throat section

INTRODUCTION

The refrigeration and air conditioning consume about 10 to 15% of the available electric energy. The depletion of fossil fuel resources and the various protocols for the protection of the environment have prompted researchers to develop cooling systems allowing the use of waste heat of industrial processes [1-2] or a free energy source, such as solar energy [3-4] and to propose solutions for improving the energy efficiency of conventional compression systems.

The aim of this work is to highlight the role of the ejector in the energy savings through two types of refrigeration systems. The first is the ejector refrigeration system which is a heat-operated system where the compressor is replaced by an ejector [5-7]. The second is the transcritical CO_2 refrigeration system where the ejector is used as the main expansion device to improve the energy efficiency of the system [8-10].

The ejector is the main component of the ejector refrigeration system. Thus, its performance is closely related to that of this component. Despite the benefits of this system including simplicity, absence of moving parts (except the existence of a boiler feed pump) and ease of implementation, its weak *COP* is the major hindrance for its broad diffusion. Further efforts are needed to increase its performance and reduce its cost.

In this paper, we review the various theoretical and experimental works carried out in the Thermal Center of Lyon (CETHIL) and the Department of Mechanical Engineering of Annaba university for few decades.

Several aspects of this study including the influence of geometrical parameters, operating conditions and the refrigerant nature on the system performance as well as the research for acceptable refrigerants in terms of environmental protection were treated.

The second part concerns the use of an ejector as an expansion device instead of an expansion valve in a transcritical CO_2 refrigeration system in order to improve its performance.

A modeling of the refrigeration system cycle including a constant-area mixing ejector has been developed. For fixed cooling capacity and ejector compression ratio, this model is used in a first step to design the ejector when it operates at transition mode. In a second step, it is used to assess the performance improvement of the ejector-expansion transcritical CO_2 system in relation to the basic system. The influence of the gas cooler pressure and the evaporator superheat on the system performance is also investigated.

Chapter I

PERFORMANCE ANALYSIS OF AN EJECTOR REFRIGERATION SYSTEM

I.1. THEORY AND OPERATING ANALYSIS OF AN EJECTOR

The ejector was invented by Sir Charles Parsons around 1901 for removing air from a steam engine condenser. In 1910, an ejector was used by Maurice Leblanc in the first steam jet refrigeration system. This system has experienced great popularity in the early 1930s for air conditioning of large buildings, before being later supplanted by systems using mechanical compressors.

In recent years, the ejector is the subject of renewed attention from researchers either as a main component of the heat-operated refrigeration systems or a booster for the refrigeration systems using mechanical compression.

The analysis of the operation of an ejector is often carried out with the one-dimensional theory [11-13]. This theory, first introduced by Keenan *et al.* [14] is based on ideal gas dynamics and the mass, momentum and energy balances. Heat and friction losses were not considered.

This model has undergone several improvements. Among them one introduced later by Munday and Bagster [15] can predict the constant-capacity of an ejector.

A typical one-dimensional flow process of the working fluid through an ejector is depicted in Fig. I.1. The high pressure fluid (P), also known as 'motive fluid' or 'primary fluid', is expanded through the convergent divergent nozzle in the ejector to produce high velocity vapor (i). It fans out with supersonic speed to create a very low pressure region at the exit plane (ii) of the primary nozzle and hence in the mixing chamber. Consequently, the higher pressure fluid, which is called 'secondary fluid' (S), can be entrained from the evaporator into the mixing chamber. The expanded motive fluid is thought to flow forming a converging duct without mixing with the driven fluid. At some cross-section along this duct, defined by Munday and Bagster as the "effective area", the secondary fluid chokes (iii). This choking occurs before mixing with the primary fluid. The choking phenomenon plays an important role in deciding the critical operational parameters of an ejector. Both fluids are expected to be completely mixed by the end of the mixing chamber. The pressure of the supersonic mixed fluid is assumed to be constant until the cylindrical part of the mixing chamber (iv). Due to a high-pressure downstream of the ejector, formation of a normal shock, that converts the flow to subsonic, is expected occurring before the fluid enters the diffuser (v). This shock wave causes a major compression effect. A further pressure raise of the flow occurs in the diffuser (vi).

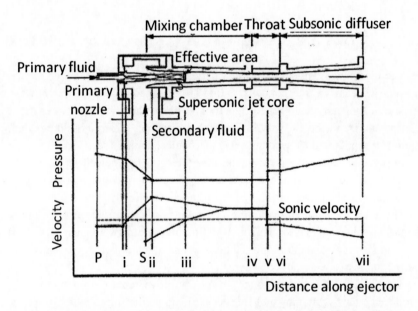

Figure I.1. Schematic view and the variation in stream pressure and velocity as a function of localization along a stream ejector [7]

According to the position of the primary nozzle, there are two types of ejector design:

- when the exit plane of the nozzle is located within the suction chamber in front of the constant-area section as described above, the mixing process is assumed to occur at constant pressure between planes 1 and 2 (Fig. I.2(a)) and the ejector is said a "constant-pressure mixing ejector";

- when the exit plane of the nozzle is located within the constant-area section, the mixing process is assumed to take place somewhere in this part of the ejector (Fig. I.2(b)) and the ejector is said a "constant-area mixing ejector".

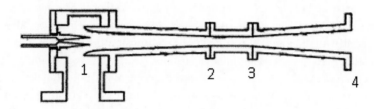

(a) "Constant-Pressure Mixing" Ejector

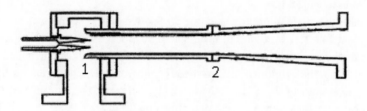

(b) "Constant-Area Mixing" Ejector

Figure I.2. Two schematic view of an ejector for modeling [7]

In the one dimensional ejector model, in general, the following assumptions are made:

1. The flow inside the ejector is steady.
2. The inner wall of the ejector is adiabatic.
3. The primary and secondary streams at the ejector inlet and the mixed flow at the ejector exit are in stagnation conditions.
4. Velocities are uniform in all sections.
5. Mixing of streams occurs at constant pressure for the "constant-pressure mixing ejector" (Fig. I.2(a)), at constant area for the "constant-area mixing ejector" (Fig. I.2(b))

6. The shock wave takes place in the cylindrical mixing chamber.
7. The working fluid is an ideal gas with a constant specific heat ratio.
8. Assumed values for the coefficients accounting for losses in the primary nozzle, the suction chamber and the diffuser are used.

In this model, the flow within the ejector is analyzed using the continuity, momentum and energy balances under the above assumptions.

Keenan *et al.* [14] have determined that the performances of constant pressure ejectors are better than those of constant-area ejectors. However, they also expressed that the theoretical results obtained by using the constant-area ejector flow model agree with experimental results, while it is difficult to obtain agreement between the theoretical and the experimental values for the constant-pressure ejector. They explained this by the difficulty to determine the ejector geometry in this last case. Other studies [16-17] suggest that the "constant-area mixing ejector" can lead to more secondary flow comparing to the "constant-pressure mixing ejector". Despite this, most studies have been focused on the constant-pressure ejector.

To make the model more realistic, the thermodynamic properties of real gases were applied [18-21]. In addition, efficiencies for the convergent-divergent, the suction chamber, the diffuser [18, 22-25] and a factor which takes into account the friction losses in the cylindrical part of the secondary nozzle have been introduced [26-27]. The values of these parameters widely differ within the range of 0.8-1, depending on ejector geometries and operating conditions [25, 28].

On the basis of experience [29-30] or by using a Computational Fluid Dynamics (CFD) software package [31-32], other researchers have recently attempted to explain the flow and mixing processes that take place through an ejector. An experimental analysis of flow visualizing in a constant-area mixing ejector with entire supersonic regime, has been carried out by Desevaux [33]. The laser induced fluorescence and laser tomography method including image processing have been used to investigate the mixing zone in the ejector.

In general, the results confirm, especially the phenomenon of the expansion wave of the motive fluid and the shock wave that occurs in the mixture, which were established by the theory of 1-D model described above.

The most important parameters which are generally used to evaluate the performance of an ejector are the entrainment ratio U and the pressure lift ratio (or compression ratio) r [5-6, 34]

$$U = \frac{\text{mass flow rate of the secondary fluid}}{\text{mass flow rate of the primary fluid}} = \frac{\dot{m}_S}{\dot{m}_P} \qquad (I.1)$$

$$r = \frac{\text{static pressure at the diffuser exit}}{\text{static pressure of the secondary flow at the suction chamber inlet}} \qquad (I.2)$$

I.2. EJECTOR REFRIGERATION SYSTEM

A schematic diagram of an ejector refrigeration system is shown in Fig. I.3. This system consists of six components in the refrigeration cycle: a vapor generator (or boiler), an ejector, a condenser, an evaporator, an expansion valve and a liquid pump. The refrigerant vapor '0' at high pressure and temperature is generated by utilizing low grade waste heat in the vapor generator. This vapor called the primary vapor enters the ejector. By expansion through a supersonic nozzle inside the ejector, supersonic flow at low pressure is formed at the nozzle exit '1'. It causes the suction of the vapor from the evaporator, called the secondary vapor, which is at state '7'. The two streams then mix with each other in the mixing chamber of the ejector. The mixed vapor at state '3' is compressed until the condenser pressure '4' by the shock wave which occurs in the constant-area section followed by flow through the diffuser. The pressurized fluid at state '4' undergoes condensation in the condenser. The condensate '5' at the condenser exit is divided into two parts: one part flows to the evaporator through an expansion valve positioned between the points '5 'and '6' and the other part returns to the pump. Since the vapor generator pressure is higher than that of the condenser, a liquid pump is used between the condenser exit '5' and the vapor generator inlet '8'. The liquid fluid entering the vapor generator is vaporized from state '8' to state '0'. The low temperature fluid in the evaporator produces a cooling effect by absorbing heat from the refrigerated medium. This is accompanied by a change of the fluid state from '6' to '7'.

The coefficient of performance of the reversible Carnot cycle for an ejector refrigeration system is defined by:

$$COP_C = \frac{T_E}{T_C - T_E} \frac{T_B - T_C}{T_B} \qquad (I.3)$$

$$COP = \frac{\dot{Q}_E}{\dot{Q}_B + \dot{W}_P} \tag{I.4}$$

where $\dot{Q}_E$ and $\dot{Q}_B$ are the heat transfer rates in the evaporator and the boiler, respectively, and $\dot{W}_P$ is the mechanical power required for pumping.

This coefficient of performance may be expressed in terms of the refrigerant thermodynamic properties:

$$COP = \frac{\dot{m}_S(h_7 - h_5)}{\dot{m}_P[(h_0 - h_8) + v_5(P_0 - P_5)]} \tag{I.5}$$

With using the entrainment ratio U defined by Eq. (I.1), the expression of COP can be written as follows:

$$COP = U \frac{(h_7 - h_5)}{[(h_0 - h_8) + v_5(P_0 - P_5)]} \tag{I.6}$$

The mechanical energy required for the pump is typically less than 1% of the heat supplied to the boiler and therefore can be neglected. In this case, the COP of the system is given by:

$$COP = U \frac{(h_7 - h_5)}{(h_0 - h_8)} \tag{I.7}$$

By assuming that the temperature of the intermediate source is equal to that of the environment and by neglecting the specific work of the pump, the system exergetic efficiency is given by:

$$\eta_{ex} = \frac{COP}{COP_C} \tag{I.8}$$

In order to make this system completely static and more competitive, two alternate ways to return the liquid refrigerant to the boiler have been proposed. The first alternative was proposed by Riffat and Holt [35]. The principle of heat pipe is applied to an ejector. The condensed liquid returns from the condenser to the boiler through a wick by capillary action. The second alternative is achieved by using the gravitational head difference between a condenser and a boiler [4, 36-37]. Thus, the requirement of the circulation pump is eliminated.

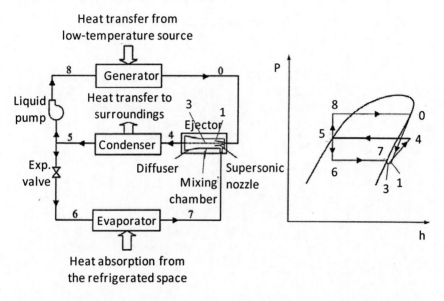

Figure I.3. A schematic diagram of an ejector refrigeration cycle and *P-h* diagram [17]

I.3. PERFORMANCE CHARACTERISTICS

The performance of an ejector refrigeration system is directly related to that of the ejector. The entrainment ratio depends on the operating conditions and the ejector geometry. The experimental studies carried out on the ejectors applied to the refrigeration [5,26,30,38] showed that their performances are limited by a critical back pressure P_C^*. For given boiler and evaporator pressures, a typical system performance curve for a fixed geometry of the ejector is shown in Fig. I.4. There are three regions: double choking flow , single choking flow and back flow.

For a condenser pressure $P_C \leq P_C^*$, the ejector entrains the same amount of secondary fluid. Therefore, the entrainment ratio, the cooling capacity and *COP* remain constant. This phenomenon is due to the secondary flow choking in the mixing chamber. When the ejector is operating in this pressure range, a transverse shock, which is responsible of a compression effect, appears in either the constant area or the diffuser section. The location of the shock wave varies with the value of the condenser back pressure. If the condenser pressure is further reduced, the shock will move toward the subsonic diffuser and vice versa. In this case the ejector is said operating in critical mode. The particular point for which $P_C = P_C^*$ called critical point corresponds to the transition mode.

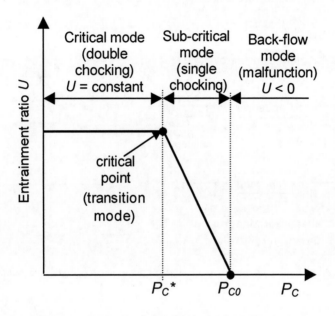

Figure I.4. Entrainment ratio versus condenser pressure for given boiler and evaporator pressures [39]

When $P_C > P_C^*$, the transverse shock tends to move backward into the mixing chamber and interferes with the mixing process of primary and secondary fluids. The secondary flow is no longer choked. Thus, the amount of secondary fluid driven by the ejector varies and the entrainment ratio begins to fall off rapidly. It is the sub-critical operating mode of the ejector. If the condenser pressure is further increased, the flow will reverse back into the evaporator and the ejector losses its function completely.

I.3.1. Influence of the Operating Conditions

Figs. I.5(a) and (b) show the results obtained by L. Boumaraf and A. Lallemand [39] for the working fluids R142b ($\Phi = 9.27$, $\varphi=6.58$) and R600a ($\Phi = 7.45$, $\varphi=6.10$) by using the 1-D model with constant-area mixing, under the assumptions described above with the isentropic efficiencies equal to 0.95 for the primary nozzle, 0.98 for the suction chamber and 0.85 for the diffuser.

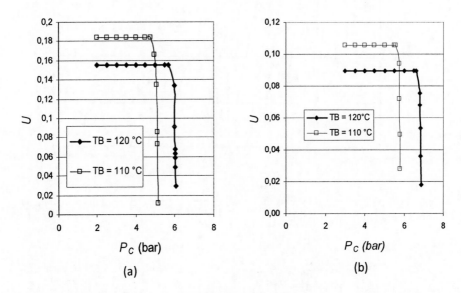

Figure I.5. Entrainment ratio versus condensation pressure obtained by modeling for two working fluids: R142b (a) and R600a (b) (Evaporator temperature: $T_E = 0°C$) [39]

These results also show that, for a fixed value of the boiler phase change temperature, the ejector entrainment ratio increases with the vaporization temperature [39].

It can be noted that for a condenser pressure below the critical value, the mixing chamber is always choked, the flow rate of the driven fluid is independent of the downstream (condenser) pressure. It can only be raised by an increase of the upstream (evaporator) pressure as shown in Fig. I.6, which is obtained from tests carried out by Lu [5] on the experimental device given in Fig. I.7.

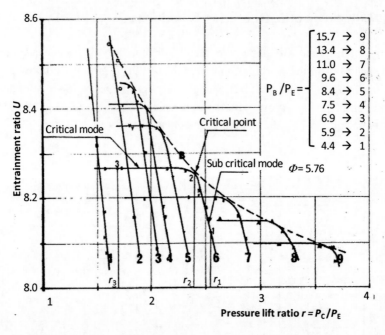

Figure I.6. Entrainment ratio versus the pressure lift ratio for a constant boiler pressure and various evaporator pressures [5]

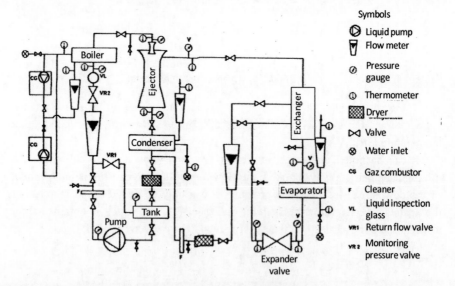

Figure I.7. Experimental device for the ejector study [5]

The critical condenser pressure is a function of the pressure and momentum of the mixed flow. The mixing pressure is close to that prevailing in the evaporator. As the secondary fluid enters the ejector at a very low speed, the momentum of the mixture depends exclusively on the primary flow at its exit from the primary nozzle. Accordingly, to increase the ejector critical pressure, the pressure at the boiler or the evaporator must be increased.

A decrease in the boiler pressure causes a decrease in the mass flow rate of the primary fluid. As the flow area in the mixing section is fixed, the secondary flow rate increases. This results in an increase of the entrainment ratio of the ejector and the cooling capacity of the system as well as of its *COP*. However, this causes a decrease in the momentum of the mixed flow. Consequently, the ejector critical pressure is reduced.

It should be noted that this increase in ejector performance depends on the temperature increase of the cold source, which is generally fixed by the operating conditions of the system.

According to the performance characteristics, for given operating conditions, the most desired ejector is that which leads to the highest entrainment ratio and maintains the highest possible discharged pressure. Therefore, it is recommended that the system operates at the critical condenser pressure.

I.3.2. Influence of the Geometry

As the operating conditions, the ejector geometry has a significant influence on the system performance [14, 16, 22, 26, 29, 40-42]. The geometrical parameters of an ejector (Fig. I.8) that appear in the literature are given in Table I.1.

According to the experimental work undertaken on the ejector, at the CETHIL [5,6,40], the only parameter of the primary nozzle that has a significant influence on the performance is the relative length of the divergent part which must be relatively small ($L_P \sim 3.5\ d$) in order to reduce the friction loss and have a better stability of the flow. In the study of Eames *et al.* [41], it was found that the influence of using small primary nozzle was similar to that of decreasing boiler saturation pressure whereas the influence of the primary nozzle exit diameter, d, was not significant. It was also established that the critical pressure of the condenser can be increased by utilizing a longer and larger mixing chamber.

Table I.1. Dimensional and dimensionless geometrical parameters of an ejector

Parameter	Definition
φ	exit / throat areas of the primary nozzle $= \left(d/d^*\right)^2$
Ω	exit / throat areas of the secondary nozzle $= \left(d_D/D\right)^2$
Φ	secondary nozzle throat area/primary nozzle throat area $= \left(D/d^*\right)^2$
(x/D)	Position of the primary nozzle relative to the mixing chamber
(L/D)	length of the mixing chamber relative to its diameter
(L_P/d)	length of the primary nozzle divergent relative to its exit diameter
(L_D/D)	length of the diffuser relative to the mixing chamber diameter
α, β, γ	characteristic angles of the primary and secondary nozzles

In Lu's work [5], the secondary nozzle consists of three parts: a convergent, a cylindrical part (mixing chamber) and a divergent (diffuser). The experimental tests have shown that the position of the primary nozzle relative to the mixing chamber, determined by the dimensionless parameter x/D (Fig. I.8), has a great influence on the ejector performance. Its optimal value is between 0.5 and 2. However, this optimum is highly dependent on the nature of the primary fluid used, specially, through its ratio of specific heats at constant pressure and constant volume. On the other hand, in order to optimize the size of the mixing chamber (cylindrical tube), a value between 8 and 12 is recommended for the geometrical parameter (L/D).

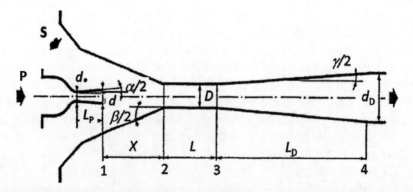

Figure I.8. Geometrical parameters of an ejector [40]

In the experimental work of Aphornratana and Eames [42], where the geometric configuration of the ejector used is different from that used by Lu, the convergent of the secondary nozzle is longer and serves as a mixing chamber (Fig. I.9). The primary nozzle is mounted on a threaded shaft which allows to adjust the distance between the nozzle exit and the mixing chamber inlet in order to determine the influence of the primary nozzle position on the performance of the ejector. It was found that, for fixed boiler and evaporator temperatures, the COP and the cooling capacity can be varied as much as 100% by changing on the nozzle position. Moving the nozzle into the mixing chamber causes the COP to fall and the cooling capacity to decrease when the boiler input is maintained constant. However, the system could be operated at a higher critical condenser pressure. Moving back the nozzle from the mixing chamber causes the cooling capacity and the COP to increase with the expense of the critical condenser pressure. According to the results of their tests, a single optimum primary nozzle position cannot be defined to meet all operating conditions of the system. It was concluded that each ejector requires a particular optimum nozzle position.

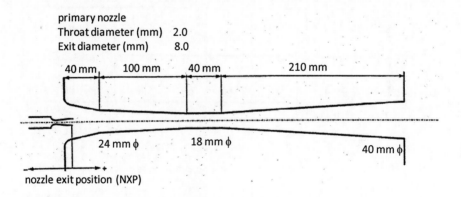

Figure I.9. Geometrical characteristics of the ejector used by [42]

However, as reported in the literature concerning the ejector, the geometrical parameter Φ is of prime importance in the operation of an ejector. Fig. I.10 shows the Nahdi's results for the refrigerant R11.

The explanations mentioned above show the influence of geometrical parameters on the ejector performance. Therefore, the constraint of a fixed geometry ejector in a refrigeration system is a major hindrance for achieving optimal performance under various operating conditions.

For a fixed evaporator temperature, Boumaraf and Lallemand [39] recommend to design the ejector (specially the geometrical parameter Φ and the cross-section of the primary nozzle throat) at a relatively high temperature of the boiler in the critical point conditions in order to guarantee a higher ejector performance in case of using at a lower temperature of the boiler with the expense of the critical condenser pressure (Fig. I.5).

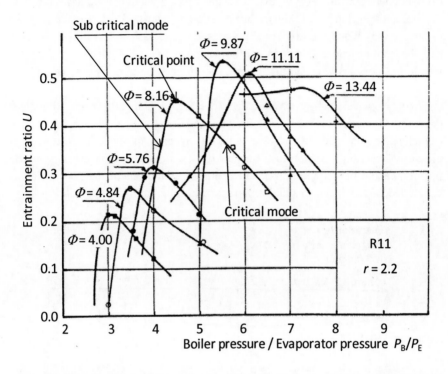

Figure I.10. Entrainment ratio versus driving pressure ratio for a constant pressure lift ratio and various values of Φ [6]

I.3.3. Empirical Correlations

According to the experimental tests carried out at the laboratory of CETHIL [5], on ejectors with different geometrical parameters Φ, operating with the refrigerant R11 in several operating conditions, it was concluded that it is possible to determine the behavior of an ejector, in any mode of operation, using a relationship between the entrainment ratio U, the pressure lift ratio r, the driving pressure ratio ξ and the geometrical parameter Φ such as:

$$f(U, \xi, r, \Phi) = 0 \tag{I.9}$$

Nadhi [6] suggests, for transition operating mode of an ejector with $\Phi < \Phi_{opt}$, (Fig. I.10) a linear relationship between the entrainment ratio U and the geometrical parameter Φ. Thus, in this operating mode, Eq. (I.9) becomes:

$$U^* = a + b\Phi f(r) \tag{I.10}$$

Moreover, it was found that this relationship is the same for the three tested CFCs: R11, R113 and R114; this was confirmed in the study of Boumaraf and Lallemand [39] on the modeling of an ejector refrigeration system operating with the refrigerants R142b and R600a. However, it is noted in this case that the evolution of the entrainment ratio U according to the lift pressure ratio in the transition mode, depends on the nature of the working fluid through the corresponding geometrical parameter of the ejector Φ, which is determined at a fixed boiler temperature and not at a fixed boiler pressure as in the previous case. We will return in more details on this point in § I.3.4.

The generalization of this observation suggests the existence of an universal function of behavior of an ejector, in transition mode.

Later, based on the experimental study of Work et $al.$ [43],[44] with two commercial types of ejectors, Dorantès and Lallemand [40] have established an empirical correlation to calculate, in transition mode the entrainment ratio as a function of the pressure lift ratio and the driving pressure ratio:

$$U^* = \left(\frac{1}{r} - 0.23\right)\left(\frac{1}{0.45\xi_{opt} - 0.06\xi_{opt}^{\ 2}}\right) \tag{I.11}$$

This relationship is valid for $2 \leq r \leq 3.5$ and $4 \leq \Phi \leq 9.87$. In this case, the geometrical parameter Φ is a linear function of the optimal driving pressure ratio ξ_{opt}, given by:

$$\Phi = -2.474 + 2.045\xi_{opt} \tag{I.12}$$

This correlation was validated against experimental data from the test campaign with the refrigerants: R11, R113 and R114. The maximum prediction error is less than 6%.

A second empirical correlation was proposed by Dorantes and Lallemand [40]:

$$U^* = 3.32\left(\frac{1}{r} - \frac{1.21}{\xi_{opt}r}\right)^{2.12}$$

(I.13)

Eq. (I.13) has the same validity range as Eq. (I.11) and the maximum prediction error is less than 5%.

By defining the operating conditions of the ejector refrigeration system, using the pressures of boiler, condenser and evaporator, the authors have established an empirical correlation which is purely dynamic, that can appear independent of the nature of the fluid.

In fact, the operating conditions of an ejector refrigeration system are determined by the temperatures of heat sources and therefore, by the phase change temperatures in the boiler, the condenser and the evaporator, obtained from heat transfer in these three heat exchangers.

Thus, for fixed operating temperatures of the system, the influence of the nature of the refrigerant must appear through a difference in saturation pressure levels in the three heat exchangers. This point will be detailed in the following section.

I.3.4. Influence of the Fluid Nature

The study of the influence of the refrigerant nature on the performance of the ejector refrigeration system that has been undertaken in the CETHIL during the 90s [45-46], was motivated by the search of potential substitutes to the CFCs: R11, R113 and R114 used as working fluids during the various campaigns of tests on this system. This substitute fluid (HCFC or HFC) is either a pure refrigerant, an azeotropic or a non-azeotropic binary mixture.

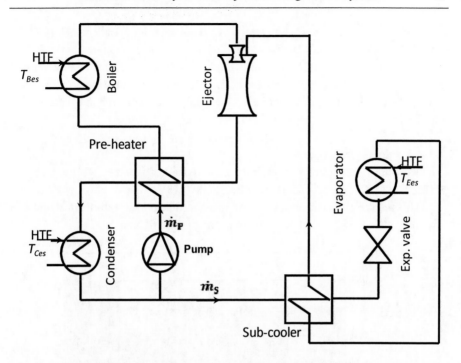

Figure I.11. A schematic diagram of an ejector refrigeration cycle with regenerators [46]

In order to optimize the system performance as a function of the working fluid, a simple model of the operating cycle of the ejector refrigeration system including two heat exchangers for energy recovery (Fig. I.11) has been developed by Dorantès and Lallemand [45]. The system performance depends mainly on the ejector entrainment ratio, and the enthalpic characteristics of the working fluid at the different points of the operating cycle. The modeling of the entrainment ratio U is obtained by assuming that the ejector always operates in the transition (optimal) mode [5]. In this case, the geometrical parameter Φ is suited for this optimal operating mode and the entrainment ratio of the ejector is evaluated using Eq. (I.13) and Eq. (I.12). The model of the thermodynamic cycle is closely related to that of the heat exchangers. The model used for heat exchangers is based on the assumption of a fixed pinch for the boiler, condenser and evaporator and also a fixed temperature difference between the inlet and the exit of each heat transfer fluids (HTF) (assumed to represent heat sources). Indeed, the working fluid temperature in each heat exchanger and consequently its pressure can be determined from the inlet temperature of HTF, the pinch and the temperature difference between the inlet and the exit and by taking into account any necessary superheating and sub-cooling.

This work has been completed by Boumaraf and Lallemand [46] by introducing a method for calculating the thermodynamic characteristics: enthalpy, entropy and specific volume of pure fluids and azeotropic and non-azeotropic binary mixtures. These properties are determined with the cubic equation of Peng Robinson and equality of fugacities of the liquid and vapor for the liquid-vapor equilibrium.

I.3.4.1. Pure Refrigerants

The pure fluids tested in this study are classical refrigerants: R11, R22, R114 R123, R133a, R134a, R141b, R142b, R152a, RC318.

The results (Fig. I.12) show that the coefficient of performance and the exergetic efficiency depend strongly on the refrigerant nature.

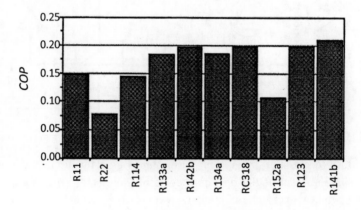

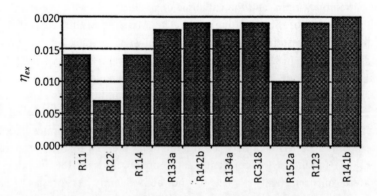

Figure I.12. *COP* and exergetic efficiency (η_{ex}) of some pure refrigerants (HTF temperatures: T_{Bes}= 90°C, T_{Ces} = 25°C, T_{Ees} = 20°C [45]

The difference between the performances of the different fluids is due mainly to the entrainment ratio (Fig. I.13) because the ratio of the enthalpy variation of the fluid in the evaporator Δh_E to that in the boiler Δh_B has a weak variation from one fluid to another (Fig. I.14). It is also noted that R141b, R142b and RC318 lead to the best system performance

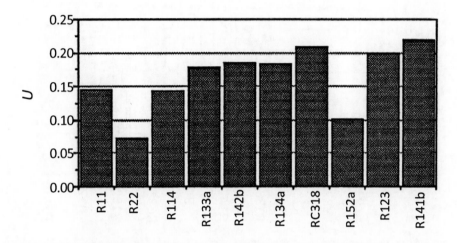

Figure I.13. Entrainment ratio of some pure refrigerants (HTF temperatures: $T_{Bes} = 90°C$, $T_{Ces} = 25°C$, $T_{Ees} = 20°C$ [45]

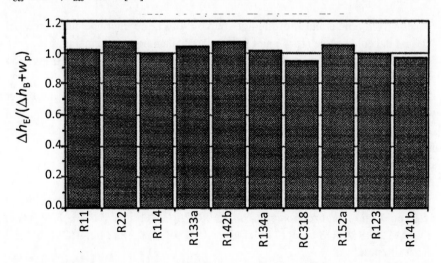

Figure I.14. Enthalpy ratio of some pure refrigerants (HTF temperatures: $T_{Bes} = 90°C$, $T_{Ces} = 25°C$, $T_{Ees} = 20°C$ [45]

By considering the temperatures of the liquid-vapor equilibrium in the boiler, condenser and evaporator, as reference for calculating the coefficient of performance of the ideal Carnot cycle, instead of the temperatures of heat sources, the exergy efficiencies are about seven to thirteen higher times (Fig. I.15).

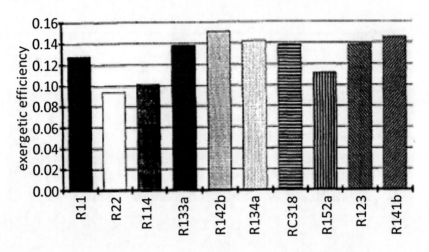

Figure I.15. Exergetic efficiency with reference to the temperatures of liquid vapor equilibrium of some pure refrigerants (HTF temperatures: $T_{Bes} = 90°C$, $T_{Ces} = 25°C$, $T_{Ees} = 20°C$ [45]

For fixed temperatures of the hot source (90°C), the intermediate source (25 °C) and the cold source (15°C) and for a reference temperature equal to that of the intermediate source, an exergetic analysis has been carried out on the refrigeration system using the fluid RC318. The results show that the major part of the exergy destruction takes place in the ejector, followed by the condenser and the boiler where the exergy losses are mainly due to the superheating of the vapor which creates great thermal differences between the refrigerant and the heat transfer fluids.

I.3.4.2. Azeotropic And Non-Azeotropic Binary Mixtures

The same model as for the pure refrigerants has been used for the binary azeotropic and non-azeotropic mixtures: R22/RC318, R22/R142b, R22/R124, R22/R152a, R22/R134a, R134a/R142b, R152a/R142b, R134a/R152 [46] and RC318/R141b [45].

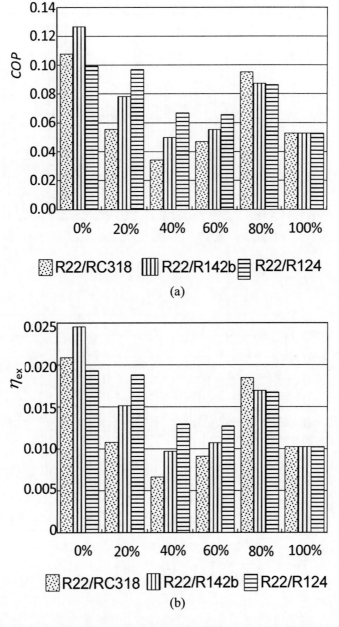

Figure I.16. Cooling efficiency (a) and second law efficiency (b) of the three strongly zeotropic mixtures as a function of R22 mole fraction [46]

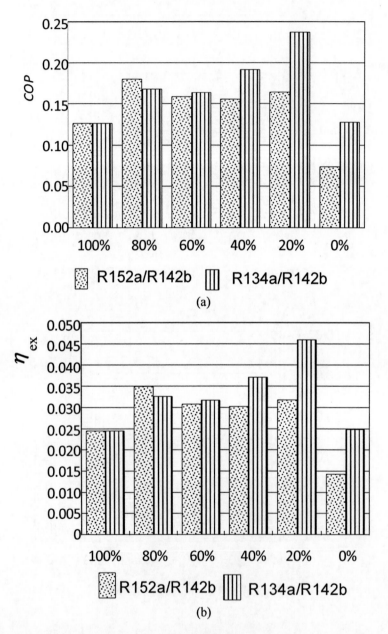

Figure I.17. Cooling efficiency (a) and second law efficiency (b) of the two moderately zeotropic mixtures as a function of R142b mole fraction [46]

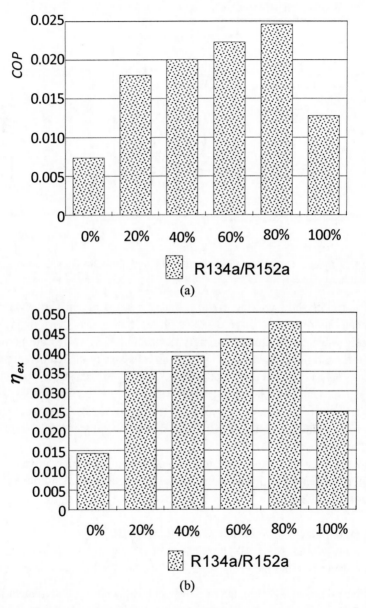

Figure I.18. Cooling efficiency (a) and second law efficiency (b) of the almost azeotropic mixture R134a/R152a as a function of R134a mole fraction [46]

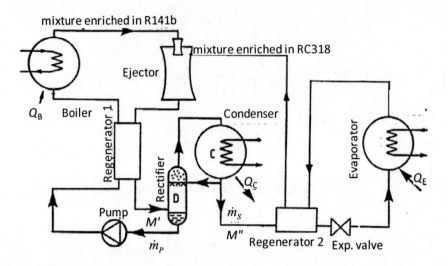

Figure I.19. Ejector system using two fluids of different molar weights. Case of the use of the mixture RC318-R141b [45]

The results show that the use of a binary mixture of refrigerants can lead to either an improvement or a degradation of the system performance. More precisely when the mixture is strongly zeotropic (R22/RC318) the performance coefficient and the exergetic efficiency (with reference to the temperatures of heat sources - HTF) decrease compared to those of one or other of the two pure fluids (Fig. I.16). On the other hand, when the mixture is moderately zeotropic (R134a/R142b) or almost azeotropic (R134a/R152a), the coefficient of performance and the exergetic efficiency increase compared to those of the two pure fluids (Figs. I.17-18).

The results also show that the use of a non-azeotropic mixture with an optimum concentration leads to the minimization of irreversibilities in the heat exchangers and ejector.

I.3.4.3. Non-Azeotropic Binary Mixture with Distillation

Various studies [43,47-48] have shown that the entrainment ratio of an ejector is improved when the molecular weight of the motive fluid is lower than that of the driven fluid. In this case, the improvement of the ejector entrainment ratio U' compared to the case where the primary and secondary fluids have the same molecular weight U is given by:

$$\frac{U'}{U} = K\left(\frac{M''}{M'}\right)^{\frac{1}{2}}$$

(I.14)

where M' and M'' are respectively the molecular weights of the primary and secondary fluids and K is a constant.

Dorantes and Lallemand [45] have investigated by using an experimental device (Fig. I.19) the behavior of the ejector refrigeration system operating with the non-azeotropic mixture R141b/RC318 separated into two flows of different molecular weights.

The simulation of the behavior of this system has been made in optimal operating mode by adopting the same assumptions as above with regard to superheating, pinching, etc.

In this case, Eq. (I.13) of the entrainment ratio of the ejector operating in transition mode has been replaced by Eq. (I.15) which was proposed by Dorantes [40] and valid for primary and secondary fluids with different molecular weights:

$$U' = \frac{1 - 1.3(1 - 1/r)}{\left[0.06\, M'/M'' + \left(0.6 - 0.12\xi_{opt}\right)\right]\xi_{opt}}$$

(I.15)

The experimental tests have focused on primary fluid composed of the mixture 0.2RC318/0.8R141b and a secondary fluid formed by the mixture 0.8RC318/0.2R141b. The ratio of the molecular weights M''/M' obtained in this case is equal to 1.37.

The results show that the addition of a rectifier at the ejector exit provides a molecular weight of the secondary fluid greater than that of the primary fluid but the system is not necessarily improved. Indeed, if the difference in molecular weights for the same values of ξ and r, must lead to an improvement of U, the characteristics of each fluid have an influence on these two parameters, what in fact leads to a reduction of U. In addition, the non-azeotropic behavior can have a positive or negative influence on the irreversibilities in the heat exchangers according to the proportions of each component in the mixture.

Chapter II

BEHAVIOR MODELING OF AN EJECTOR REFRIGERATION SYSTEM IN "DESIGN" AND "OFF-DESIGN" CONDITIONS

In order to evaluate more precisely the performance and the characteristics of the operating cycle of an ejector refrigeration system for air conditioning (Fig. II.1), a relatively fine modeling of the overall system, based on its components models has been developed [39]. The cycle is determined using the temperatures of the three heat sources and local heat transfer coefficients for the boiler, the condenser and the evaporator. In addition, this simulation program includes a correlation of the ejector entrainment ratio established in transition mode. At first the ejector is designed in transition mode using the balance equations of the 1-D model described above for a cooling capacity of 10 kW and fixed temperatures of the hot, intermediate and cold sources. Two examples are treated:

- in the first case, the operation of the ejector is analyzed using the constant-area ejector flow model with the working fluids R142b and R600a assumed as an ideal gas;
- the second example concerns a two-phase ejector which operates with the working fluid R717. In this case, a constant-pressure ejector flow model is used for the analysis.

Thereafter the other system components are designed when the ejector operates in the transition mode in the same conditions with the fluids R142b and R600a. When the system is operating in these conditions, it is said to operate "in design conditions" and when it is operating out of these conditions, it is said to operate "off-design conditions".

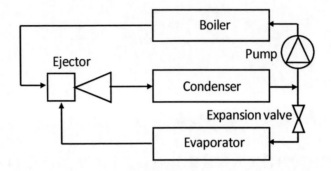

Figure II.1. Schematic view of a jet refrigeration system [39]

II.1. MODELING OF THE EJECTOR BEHAVIOR

II.1.1. Ideal Gas: Case of R142b and R600a

- Assumptions and useful equations

The ejector flow model considered is the constant-area mixing model (Fig. II.2) with the assumptions mentioned in §I.1 and the same values of the isentropic efficiency for the primary nozzle, the suction chamber and the diffuser as those given in § I.3.1.

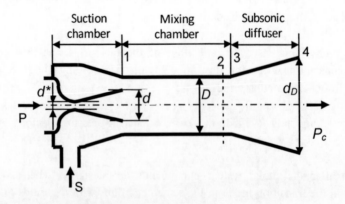

Figure II.2. Schematic diagram of the ejector for the constant-area model [39]

The model governing equations are obtained by applying mass, momentum and energy balances across the respective control volumes [13,23, 26-27].

Velocity V and mass flow rate $\dot{m}$ of a fluid at any cross-section A of a convergent-divergent nozzle can be expressed as:

$$V = \sqrt{2\eta(h_0 - h)}_{\text{is}} \tag{II.1}$$

$$\dot{m} = \rho A\sqrt{2\eta(h_0 - h)}_{\text{is}} \tag{II.2}$$

where h is the perfect gas enthalpy given by:

$$h = c_{\text{p}}T \tag{II.3}$$

Primary flow choking occurs at the cross-section throat A_{P}^{*} thus, its mass flow rate is obtained by:

$$\dot{m}_{\text{P}} = \frac{P_{\text{B}}A_{\text{P}}^{*}}{\sqrt{T_{\text{P0}}}} \times \sqrt{\frac{\gamma}{R}\left(\frac{2}{\gamma+1}\right)^{(\gamma+1)/(\gamma-1)}} \sqrt{\eta_{\text{N}}} \tag{II.4}$$

For the critical mode, secondary flow choking occurs before mixing with the primary flow, so its mass flow rate can be calculated at the cross-section throat A_{S1} by:

$$\dot{m}_{\text{S}} = \frac{P_{\text{E}}A_{\text{S1}}}{\sqrt{T_{\text{S0}}}} \times \sqrt{\frac{\gamma}{R}\left(\frac{2}{\gamma+1}\right)^{(\gamma+1)/(\gamma-1)}} \sqrt{\eta_{\text{S}}} \tag{II.5}$$

By applying momentum, energy and mass balances between the inlet section of the cylindrical mixing chamber and the section before the shock wave, the mixture velocity V_2, temperature T_2 and pressure P_2 are given by the following equations:

$$V_2 = \frac{V_{P1} + U V_{S1}}{1+U} + \frac{P_{P1} A_{P1} + P_{S1} A_{S1} - P_2 (A_{S1} + A_{P1})}{(1+U) \dot{m}_P} \tag{II.6}$$

$$\left(C_P T_{P1} + \frac{V_{P1}^2}{2} \right) + U \left(C_P T_{S1} + \frac{V_{S1}^2}{2} \right) = (1+U) \left(C_P T_2 + \frac{V_2^2}{2} \right) \tag{II.7}$$

$$\dot{m} = \dot{m}_P + \dot{m}_S = \frac{P_2}{R T_2} (A_{P1} + A_{S1}) V_2 \tag{II.8}$$

The Mach number M_2 of the mixed flow can be evaluated from:

$$M_2 = \frac{V_2}{\sqrt{\gamma R T_2}} \tag{II.9}$$

The Mach number of the fluid after the shock wave M_3 and the corresponding pressure lift ratio P_3/P_2 are expressed as:

$$M_3 = \sqrt{\frac{1 + [(\gamma - 1)/2] M_2^2}{\gamma M_2^2 - [(\gamma - 1)/2]}} \tag{II.10}$$

$$\frac{P_3}{P_2} = 1 + \frac{2\gamma}{\gamma + 1} (M_2^2 - 1) \tag{II.11}$$

The temperature T_4 and the pressure P_{4cal} of the compressed fluid at the diffuser exit can be derived from:

$$h_4 = \frac{h_{P0} + U h_{S0}}{1+U} \tag{II.12}$$

$$\eta_D = \frac{h_3 - h_{4is}}{h_3 - h_4} \tag{II.13}$$

The cooling capacity is given by:

$$\dot{Q}_E = \dot{m}_S \Delta h_E \qquad \text{(II.14)}$$

In transition mode at the primary nozzle plane 1, where the two flows first meet, the static pressure is assumed to be uniform and determined by the choking condition of the secondary flow in this plane. The mixing of the two flows is supposed to be complete before the normal shock wave which occurs at the end of the cylindrical mixing chamber. Besides, the model uses superheat at the evaporator exit $\Delta T_E = 5\,\text{K}$ (in the case of R142b), $\Delta T_E = 3\text{K}$, (in the case of R600a) and at the boiler exit $\Delta T_B = 3\,\text{K}$ in both cases. These superheating values allow to realize dry expansions in the primary and the secondary nozzles (i.e. that the temperature at the end of expansion is higher than its saturation temperature in the final state).

- Design method of the ejector

The model described above is applied to the design of an ejector operating in transition mode. For fixed values of phase change temperatures of the boiler T_B, the condenser T_C and the evaporator T_E, the driven mass flow rate, $\dot{m}_S$, is fixed for a cooling capacity of 10 kW (Eq. (II.14)) and the cross-section A_{S1} can then be deduced from Eq. (II.5). For any $\dot{m}_P$ value, A_P^* and A_{P1} can be calculated using Eq. (II.4) and Eq. (II.2), respectively and consequently A_1 ($A_1 = A_{P1} + A_{S1}$). The resolution of the balance equations in the mixing tube and the diffuser (Eqs. (II.6-II.13)) leads to a new value of P_{4cal}, which will be adjusted by changing $\dot{m}_P$ until the equality with P_C. The output parameters are the cross-section area A_P^*, the geometrical parameters φ and Φ, the mass flow rate $\dot{m}_P$ and the entrainment ratio U^*.

For condensation and evaporation temperatures equal to 45 and 0°C, respectively, Figs. II.3, 4 and 5 depict the effect of the boiler temperature on U^*, φ, Φ and A_P^* in the selected operating range for R142b and R600a. In these cases, the ejector geometry is adapted for each operating conditions in order to optimize its performance. Thus, it is noted that U^* increases with T_B and that R142b leads to better results than R600a.

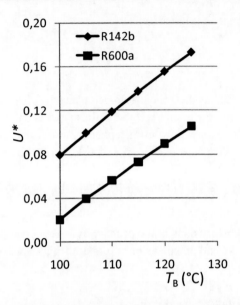

Figure II.3. Variation of the entrainment ratio at the critical point according to the boiler phase change temperature (with T_C=45°C and T_E=0°C) for several ejector geometries [39]

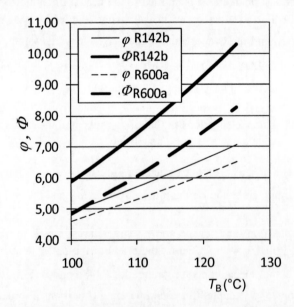

Figure II.4. Variation of the ejector geometrical parameters according to the boiler phase change temperature (with T_C=45°C and T_E=0°C) at the critical point [39]

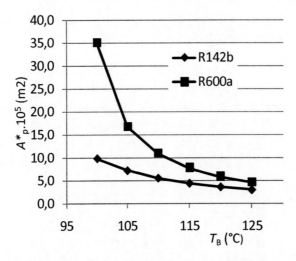

Figure II.5. Variation of the primary nozzle throat section area according to the boiler phase change temperature (with T_C=45°C and T_E=0°C) at the critical point [39]

- Evolution of the performance of an ejector with fixed geometry

The previous equations are then applied to the research of the various critical points of the operation of four ejectors which have been sized according to the procedure described above at T_B=120°C and 110°C. The condenser and the evaporator phase change temperatures are fixed at 45°C and 0°C, respectively. These ejectors are noted EJ1 (T_B=120°C), EJ2 (T_B=110°C) for R142b and EJ3 (T_B=120°C), EJ4 (T_B=110°C) for R600a in Table II.1. For several values of T_B (90≤T_B≤125°C) and T_E (0≤ T_E≤15°C), the critical condenser pressure P_C^* and the optimal entrainment ratio U^* corresponding to a secondary flow choking located at the inlet of the mixing chamber (plane 1) are calculated. Fig. II.6 shows how the optimal entrainment ratio U^* varies with the critical pressure lift ratio (r^* =P_C^*/P_E) for several values of the expansion ratio (Γ=P_B/P_E). It is noted that these variations do not depend explicitly on Γ, P_B or P_E. All theses curves have as an equation:

$$U^* = A\left(r^*\right)^{-a} \tag{II.15}$$

A, a are parameters which depend on the fluid nature and the ejector geometry (Table II.1). In fact, the variation of a with the nature of the fluid and especially with the ejector geometry is weak.

Contrarily to the results given in (Fig. II.3), with T_E constant, U^* in this case decreases with T_B. More precisely, U^* increases when the value of T_B is lower than that corresponding to the design value and decreases when it is higher than this value. This is due to the fact that P_C^* increases with T_B what increases r^* and consequently made fall U^*. In addition, at fixed T_B, U^* increases with T_E, what is in conformity with the design case.

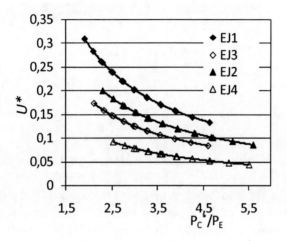

Figure II.6. Entrainment ratio of 4 ejectors operating at the critical point with R142b and R600a [39]

Table II.1. Characteristics of the four ejectors: $T_C=45°C$, $T_E=0°C$

Ejector	EJ1	EJ2	EJ3	EJ4
Refrigerant	R142b	R142b	R600a	R600a
T_B (°C)	120	110	120	110
Φ	9.27	7.45	7.45	6.04
φ	6.58	5.71	6.10	5.31
A	0.573	0.439	0.351	0.221
a	-0.957	-0.951	-0.948	-0.943

By using the 1-D model, for fixed evaporator temperature $T_E=0°C$ and boiler temperatures $T_B=110$ and $120°C$, Fig. I.5 shows the variations of the entrainment ratio U with the back pressure P_C of the three operating modes (critical, sub-critical and back flow) for the ejectors EJ1 (Fig. I.5(a)) and EJ3 (Fig. I.5(b)) respectively.

The critical mode concept ($P_C < P_C^*$) of an ejector must be specified. Indeed, for fixed cold temperature, the evolution of its performance according to the hot temperature is not similar according to whether the geometry is fixed (in this case U^* decreases when T_B increases) or adapted (in this case U^* increases when T_B increases) in spite of the assumption of a secondary flow choking at the inlet of the mixing chamber in both cases. So, for given T_C and T_E temperatures, it would be interesting to dimension the system for a relatively high temperature of the hot source in order to guarantee to the ejector an appreciable performance in the case of an operating at a lower temperature of the hot source but this, is of course with the detriment of a lower critical back pressure.

II.1.2. Real Gas: Case of R717

- Assumptions and useful equations

The ejector flow model is the constant-pressure mixing model (Fig. II.7) with the same assumptions as those mentioned in §I.1. except that relating to the state of ideal gas, which is no longer valid. The isentropic efficiency values for the primary nozzle, the suction chamber and the diffuser are also the same as those given in § I.3.1.

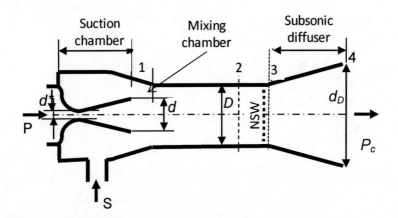

Figure II.7. Schematic diagram of the ejector for the constant-pressure model

The thermodynamic characteristics of ammonia (R717) are calculated by the REFPROP® package.

Assuming that the pressure before mixing is P_1, the following equations for the ejector section before mixing can be identified. The entropy s_{P0} and the enthalpy h_{P0} of the primary fluid at the ejector inlet are given by:

$$s_{P0} = f(T_B + \Delta T_B, P_B) \tag{II.16}$$

$$h_{P0} = f(T_B + \Delta T_B, P_B) \tag{II.17}$$

The primary stream is accelerated as its pressure drops from P_B to P_P^* corresponding to its choking, which occurs at the throat section of the nozzle. An isentropic expansion process is used to determine the final state:

$$s_{P,is}^* = s_{P0} \tag{II.18}$$

The corresponding enthalpy of the motive stream at the end of the isentropic expansion process $h_{P,is}^*$ can be determined from the state equation:

$$h_{P,is}^* = f(s_{P,is}^*, P_P^*) \tag{II.19}$$

Using the definition of expansion efficiency, the actual enthalpy h_P^* of the motive stream at the throat section of the primary nozzle can be found:

$$\eta_P^* = \frac{h_{P0} - h_P^*}{h_{P0} - h_{P,is}^*} \tag{II.20}$$

The sound speed a_P^* is derived from the following expression computed by making small variations in local values of the pressure and density:

$$a_P^* = \sqrt{\left(\frac{\Delta P}{\Delta \rho}\right)_{P,\text{is}}^*}$$

(II.21)

Using the conservation of mass, the area cross-section A_P^* of the primary nozzle throat is determined:

$$A_P^* = \frac{\dot{m}_P v_P^*}{a_P^*}$$

(II.22)

where v_P^* is the local specific volume of the motive fluid, which can be found by a property relationship:

$$v_P^* = f(P_P^*, h_P^*)$$

(II.23)

The expansion of the primary flow continues in the primary nozzle divergent corresponding to a drop in pressure from P_P^* to P_1. The fluid characteristics at the end of this expansion are determined according to the same procedure as that of the convergent.

Particularly, the Mach number M_{P1}, the cross-section area of the primary nozzle exit A_{P1} and the geometrical parameter φ are given by:

$$M_{P1} = \frac{V_{P1}}{\sqrt{\left(\frac{\Delta P}{\Delta \rho}\right)_{P1,\text{is}}}}$$

(II.24)

$$A_{P1} = \frac{\dot{m}_P v_{P1}}{V_{P1}}$$

(II.25)

$$\varphi = \frac{A_{P1}}{A_P^*}$$

(II.26)

where V_{P1} is the primary flow velocity in plane 1 calculated by Eq. (II.1) and v_{P1} is the corresponding specific volume determined from the pressure P_1 and the enthalpy h_{P1} (Eq. (II.23)).

The process of the expansion of the secondary fluid in the suction chamber is similar to that of the motive fluid in the primary nozzle convergent. In transition mode, the secondary flow choking occurs before mixing with the primary flow somewhere between planes 1 and 2, so, the throat section area can be calculated from:

$$A_S^* = \frac{\dot{m}_S \times v_S^*}{a_S^*} \tag{II.27}$$

where $\dot{m}_S$ is the mass flow rate of the secondary fluid, calculated from Eq. (II.14), a_S^* is the local sound velocity determined by an equation similar to Eq. (II.21) for the primary fluid and v_S^* is the corresponding specific volume determined from the pressure P_S^* and the enthalpy h_S^* (Eq. (II.23)).

Mixing of both primary and secondary flows is supposed to take place somewhere between the sections A_1- plane 1 and A_2-plane 2 at constant pressure P_1. The velocity V_2, the enthalpy h_2 and the cross-section area of the cylindrical tube A_2 are calculated from balance equations of mass, momentum and energy applied to the mixture between A_1 and A_2 for any fixed value of the primary flow:

$$V_2 = \frac{V_{P1} + U V_{S1}}{1 + U} \tag{II.28}$$

$$h_{P1} + \frac{V_{P1}^2}{2} + U \left(h_{S1} + \frac{V_{S1}^2}{2} \right) = (1 + U) \left(h_2 + \frac{V_2^2}{2} \right) \tag{II.29}$$

$$\dot{m} = \frac{V_2 \times A_2}{v_2} = \dot{m}_S + \dot{m}_P \tag{II.30}$$

with:

$$v_2 = f(P_1, h_2) \tag{II.31}$$

$$\Phi = \frac{A_2}{A_P^*} \tag{II.32}$$

The shock wave is assumed to occur in the cylindrical tube before entering the diffuser. The mixture characteristics downstream of the shock wave (plane 3), P_3, h_3, V_3, are also calculated from the equations of balance applied between sections A_2 and A_3:

$$\dot{m}V_2 + P_1 A_2 = \dot{m}V_3 + P_3 A_2 \tag{II.33}$$

$$\left(h_2 + \frac{V_2^{\,2}}{2} \right) = \left(h_3 + \frac{V_3^{\,2}}{2} \right) \tag{II.34}$$

$$\dot{m} = \frac{V_3 \times A_2}{v_3} = \dot{m}_S + \dot{m}_P \tag{II.35}$$

with:

$$v_3 = f(P_3, h_3) \tag{II.36}$$

After the shock wave, the fluid is compressed again in the diffuser. The pressure P_{4cal} and the enthalpy h_4 at the output are derived from the following equations:

$$s_{4,is} = s_3 \tag{II.37}$$

with :

$$s_3 = f(P_3, h_3) \tag{II.38}$$

$$h_4 = \frac{h_{P0} + U h_{S0}}{1 + U} \tag{II.39}$$

$$\eta_D = \frac{h_{4,is} - h_3}{h_4 - h_3}$$ (II.40)

$$P_{4cal} = f(h_{4,is}, s_{4,is})$$ (II.41)

$$x_4 = f(P_4, h_4)$$ (II.42)

In transition mode, in addition to the previous assumptions, the mixing constant pressure P_1 is assumed to be uniform and determined by the secondary flow choking. It means that $P_{P1} = P_{S1} = P_1 = P_S^*$ between A_1 and A_2. The mixing of the two flows is supposed to be complete before the normal shock wave.

- Design method of the ejector

The model described above is applied for designing a two-phase ejector for air conditioning application operating in transition mode with the working fluid R717 (Fig. II.8).

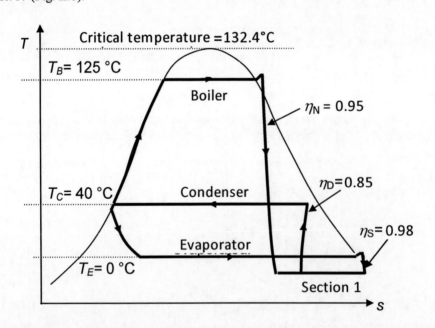

Figure II.8. *T,s* diagram of R717 ejector refrigeration cycle

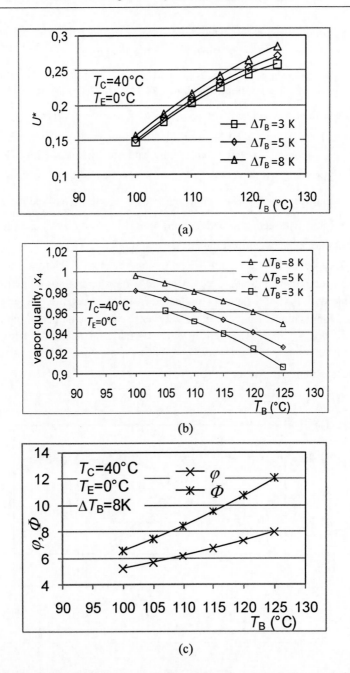

(a)

(b)

(c)

Figure II.9. Ejector characteristics versus boiling temperature in transition mode for a refrigeration system with ammonia

In this model the phase change temperatures of the boiler T_B, the condenser T_C and the evaporator T_E, the superheat at the evaporator exit $\Delta T_E = 5\,K$ and several superheat values at the boiler exit ΔT_B have been used. The driven mass flow rate is fixed for a cooling capacity of 10 kW (Eq. (II.14)), the cross-section area A_S^* can then be deduced from Eq. (II.27). For any $\dot{m}_P$ value, A_P^* and A_{P1} can be calculated using Eq. (II.22) and Eq. (II.25), respectively. The resolution of the balance equations in the mixing tube and the diffuser (Eqs. (II.28-42)), respectively leads to a new value of P_{4cal}, which will be adjusted by changing $\dot{m}_P$ until the equality with P_C. The output parameters are the cross-section area A_P^*, the geometrical parameters φ and Φ, the mass flow rate $\dot{m}_P$, the entrainment ratio U^* and the vapor quality at the ejector exit x_4.

For $T_C = 40°C$, $T_E = 0°C$ and $\Delta T_B = 3$, 5 and 8K Fig. II.9 shows the effect of the boiler temperature on U^*, x_4, φ and Φ in a operating range for R717 compatible with that of a flat solar collector. In these cases, the ejector geometry is adapted for each operating condition in order to optimize its performance. Thus, it is noted that U^* increases with T_B (Fig. II.9(a)) and decreases with T_C (Fig. II.10). It is also noted that the vapor quality at the ejector exit decreases with T_B and increases with ΔT_B (Fig. II.9(b)).

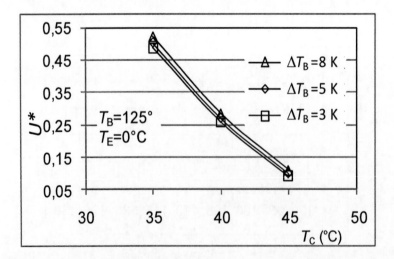

Figure II.10. Ejector characteristics versus condensation temperature in transition mode for a refrigeration system with ammonia

II.2. Designing and Modeling Behavior of an Ejector Refrigeration System with the Working Fluids R142b and R600a

The modeling of the ejector refrigeration system is based on those of its various components, which in addition to the ejector are mainly the two-phase heat exchangers (boiler, condenser and evaporator).

At first, the three heat exchangers surfaces are dimensioned in the same conditions as the ejectors: EJ1, EJ2 for R142b and EJ3, EJ4 for R600a, with a liquid sub-cooling at the condenser exit assumed equal to 3K. The corresponding flow rate values of the Heat Transfer Fluids (HTF) in the boiler, the condenser and the evaporator are then deduced from this calculation.

For a fixed cold source temperature, the model of the ejector refrigeration system is used to evaluate its performance when it is operating in design conditions and then its *COP* variation according to the hot source temperature out of these operating conditions is studied. The heat source temperatures are assumed equal to those of the HTF at the inlet of each corresponding heat exchanger.

II.2.1. Design of the Different Components of an Ejector Refrigeration System

In order to determine the heat transfer surfaces of the boiler, the condenser and the evaporator, an overall heat transfer coefficient is considered for the one-phase zones of the refrigerant and the HTF whereas for the two-phase zones, local heat transfer coefficients are used. The HTF and refrigerant flow regimes are supposed to be turbulent with $Re_L \geq 10^5$ and $Re_V \geq 10^6$.

The condenser is a double pipe heat exchanger. The HTF circulates in the inner tube and the refrigerant in the annular space in the opposite direction. In this case, three heat transfer zones are considered: desuperheating, condensation and sub-cooling.

The Dittus and Boelter correlation with a Prandtl number exponent equal to 0.4 for heating and 0.3 for cooling [49] is used for computing the Nusselt number of the HTF and the refrigerant in the one-phase zones and the Nusselt correlation [50] in the condensation zone.

The pinch, located at the vapor-saturated point for the condenser, is supposed equal to 5K. The HTF exit temperature t_{sC} and its flow rate $\dot{M}_{wC}$ can be calculated by thermal power balances on various heat transfer zones from its inlet temperature t_{eC}, the phase change temperature T_C and the sub-cooling ΔT_σ (fixed) and using the ejector program calculating the mass flow rate $\dot{m}$ and the refrigerant enthalpy at the condenser inlet (or the ejector exit) h_{eC}.

The refrigerant Reynolds number in the desuperheating zone is fixed at 10^6. The internal and external tube diameters can be calculated in such a way that the HTF Reynolds number is $\geq 10^5$. Consequently, the heat transfer surfaces of the one-phase zones can be calculated using the method of the logarithmic mean temperature difference, whereas that of the two-phase zone is determined by incrementing the vapor quality x starting from the sub-cooling zone exit according to the following equations:

$$S_C = \sum_i \Delta S_C = \sum_i \frac{\dot{m}L_V \Delta x}{\alpha_C (T_C - T_{Pi})} \tag{II.43}$$

with :

$$T_{Pi} = f(t_i) \tag{II.44}$$

and

$$t_{i+1} = t_i + \frac{\dot{m}L_V \Delta x}{\dot{M}_{wC} C_w} \tag{II.45}$$

In these calculations, the thermodynamic characteristics (h, ρ) of the refrigerant in the liquid and vapor phases are obtained using a method developed before [46] and the transport properties (λ, μ, σ) are calculated using numerical correlations depending on the temperature.

The evaporator and boiler models are similar except the existence in more for the boiler of a liquid heating zone while the fluid penetrates in the evaporator with $x > 0$, undergoes an evaporation followed by a superheating. Moreover, as the motive flow is more important than the driven flow, the boiler is made of three tubes placed in a tube of a bigger diameter whereas the evaporator consists of two coaxial tubes. In both cases, refrigerant and HTF circulate in opposite directions. The Dittus and Boelter correlation is used under the same conditions as those of the condenser and the Chen correlation [51] is applied to the two-phase zones of the boiler and the evaporator. The pinch, located at the liquid saturated point for the boiler and at the inlet for the evaporator, is assumed equal to 5K. In each heat exchanger, the HTF exit temperature and its mass flow rate can be calculated by heat balances carried out in the heat exchanger from its inlet temperature, the refrigerant phase change temperature and the corresponding superheat (fixed) and using the ejector program calculating the motive (boiler) or the driven (evaporator) refrigerant mass flow rate.

The design process is the same as in the case of the condenser, except the existence of transition zones vapor $(0.9 < x < 1)$ and liquid $(0 < x < 0.1)$ in the boiler and only vapor in the evaporator, where the correlation of Dittus & Boelter is applied by means of thermodynamic characteristics of the liquid and vapor at the corresponding saturation temperature.

In this study, the circulation pump efficiency is assumed equal to 50% and then the pump power is calculated using the equation:

$$\dot{W}_P = 2\dot{m}_P \frac{P_B - P_C}{\rho_L} \tag{II.46}$$

The refrigerant is considered mostly located in the two-phase heat exchangers (boiler, condenser and evaporator). So, its mass in the system C is evaluated using a method described by Rigot [52], which is based on the filling amount Θ of the two-phase heat exchanger. The principal stages of this method are given below:

$$C = W\left(\rho_L \, \Theta_L + \rho_V \, \Theta_V\right) \tag{II.47}$$

$$\Theta_L + \Theta_V = 1 \text{ and}$$

$$\Theta_L = \frac{B}{(B-1)^2} \ln B - \frac{1}{(B-1)} \tag{II.48}$$

in the case of the condensation and

$$\Theta_L = \frac{1}{(B-1)^2}\left[\frac{B}{(1-x_{sE})} \ln \frac{B}{x_{sE}(B-1)+1} - (B-1)\right] \tag{II.49}$$

in the case of the evaporation with :

$$B = \rho_L V_L \Big/ \rho_V V_V \tag{II.50}$$

Rigot recommends using V_L/V_V equal to the unity for the evaporation and 0.5 for the condensation.

For a cooling capacity of 10 kW, temperatures of condensation T_C=45°C (corresponding to HTF temperature at the condenser inlet t_{eC}=35°C) and evaporation T_E=0°C (corresponding to HTF temperature at the evaporator inlet t_{eE}=10°C) and for two values of the boiling temperature T_B=120 and 110°C (corresponding to HTF temperatures at the boiler inlet t_{eB}=130 and 120°C, respectively), the boiler, condenser and evaporator heat transfer surfaces are evaluated in the same conditions as the ejectors (EJ1, EJ2 EJ3, EJ4), which have been designed above for the transition mode. The corresponding ejector refrigeration systems are noted ERS1, ERS2, ERS3, ERS4, respectively in Table II.2. The main characteristics of these systems operating in their design conditions are given in Table II.2. For more details on the design of the different components of an ejector refrigeration system, to refer to [53].

Table II.2. Characteristics of the ejector refrigeration systems operating in the design conditions at transition mode ($\dot{Q}_E$ = 10 kW, t_{eE} = 10°C and t_{eC} = 35°C).

System	t_{eB} (°C)	S_E x 100 (m²)	S_B x 10 (m²)	S_C (m²)	$\dot{M}_{wC}$ (kg/s)	$\dot{M}_{wB}$ (kg/s)	Φ	φ	A_P^* x 10 (cm²)	U x 100	COP x 100	$\dfrac{COP}{COP_C}$ x 100[a]	$\dfrac{COP}{COP_C}$ x 100[b]	C (kg)
ERS1	130	3.06	6.53	7.19	4.08	2.12	9.27	6.58	3.55	15.5	10.5	3.95	9.10	11.8
ERS2	120	3.06	19.7	9.03	5.17	3.21	7.45	5.71	5.50	11.9	8.91	3.64	8.65	11.8
ERS3	130	2.24	25.8	12.4	7.09	3.78	7.45	6.10	5.82	8.96	5.77	2.16	4.99	9.66
ERS4	120	2.24	92.8	19.6	10.9	6.95	6.04	5.31	10.9	5.62	3.69	1.51	3.59	21.9

a: COP_C calculated with HTF temperatures.
b: COP_C calculated with refrigerant temperatures.

II.2.2. Behavior Modeling of the System Operating in Design and Off-Design Conditions

For an ejector refrigeration system with fixed ejector geometry and heat transfer surfaces of boiler, condenser and evaporator, the operating cycle and its characteristics can be determined for any operating conditions (HTF mass flow rates and their temperatures at the three heat exchangers inlet) using the calculation procedure detailed in Fig. II.11.

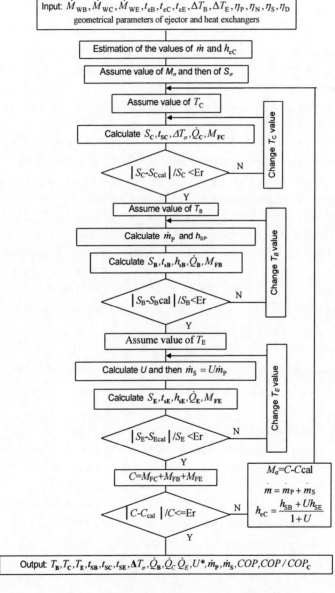

Figure II.11. Flowchart of the operating simulation program of the ejector refrigeration system [39]

The iteration sequence includes mainly the following steps.

1. Define the geometrical parameters of the ejector (A_P^*, Φ, φ) and the heat transfer surfaces of boiler (S_B), condenser (S_C) and evaporator (S_E).

2. Define the HTF mass flow rates ($\dot{M}_{wB}, \dot{M}_{wC}, \dot{M}_{wE}$) and its inlet temperatures (t_{eB}, t_{eC}, t_{eE}) in the three heat exchangers.

3. Define the refrigerant superheat values at the boiler exit ΔT_B and the evaporator exit ΔT_E and the isentropic efficiencies of the primary nozzle η_N, the suction chamber η_S and the diffuser η_D as well as the pump mechanical efficiency η_P.

4. An estimate is made for the refrigerant mass flow rate in the condenser $\dot{m}$ and its enthalpy at the ejector exit (or the condenser inlet) h_{eC}.

5. An estimate is made for the refrigerant mass M_σ and consequently that of the heat transfer surface of the sub-cooling zone S_σ.

6. An estimate is made for the condensation temperature T_C.

7. The condenser program calculates the temperature T_C by iteration until obtaining the equality between the design value of S_C and that calculated by the model. Then, the mass M_{FC} and the sub-cooling value ΔT_σ (or the refrigerant enthalpy h_{sC} at the exit) of the refrigerant in the condenser are determined. The HTF temperature t_{sC} and the heat transfer rate $\dot{Q}_C$ are also calculated.

8. An estimate is made for the boiling temperature T_B.

9. Then, the primary mass flow rate $\dot{m}_P$ is calculated by Eq. (II.4) and consequently the refrigerant enthalpy at the exit of the pump h_{sP} by Eq. (II.46).

10. The heat transfer surface of the boiler S_B is calculated using the boiler program and compared with the design value. From there, a new value of T_B is estimated and the previous step is repeated until reaching the desired value of S_B. Therefore, the refrigerant mass in the boiler M_{FB} and its enthalpy at the exit h_{sB} are calculated as well as the HTF temperature t_{sB} and the heat transfer rate $\dot{Q}_B$.

11. An estimate value of T_E is made.

12. U is calculated for the critical mode with $P_C < P_C^*$ and the critical point (U^*) with $P_C = P_C^*$ from Eq. (II.15) (since $U = U^* = $ cste for $P_C <= P_C^*$) and consequently the mass flow rate $\dot{m}_S$ of the driven fluid is calculated too.

13. The heat transfer surface S_E is evaluated using the evaporator program and its comparison with the design value allows estimating a new value of T_E and the preceding stage is repeated until equality between design and calculated values of S_E. So, the refrigerant mass in the evaporator M_{FE} and its enthalpy at the exit h_{sE} are calculated as well as the exit HTF temperature t_{sE} and the cooling capacity $\dot{Q}_E^*$.

14. Lastly, the total mass of the refrigerant in the system C is calculated and then compared with the design value. From there, a new value of the refrigerant mass in the sub-cooling zone M_σ is estimated and then that of the heat transfer surface S_σ. New values of the mass flow rate of the refrigerant $\dot{m}$ and its enthalpy at the condenser inlet h_{eC} are also obtained.

15. The preceding stages of calculation ((6)-(14)) are repeated until obtaining the equality between the design and the calculated values of the refrigerant mass in the system C.

The operating characteristics of the system: T_B, T_C, T_E, $t_{sB,}$ $t_{sC,}$ t_{sE}, ΔT_σ, $\dot{Q}_B$, $\dot{Q}_C$, $\dot{Q}_E$, U^*, $\dot{m}_P$, $\dot{m}_S$, COP, COP/COP_C are at the output of this simulation program.

For three values of the hot source temperature, which are below the design value $t_{eB}=120°C$, equal to the design value $t_{eB}=130°C$ and above the design value $t_{eB}=135°C$, respectively and for a fixed cold source temperature $t_{eE} = 10°C$, Fig. II.12 shows the effect of the intermediate source temperature on the COP of ERS1 and ERS3 systems in a selected operating range, corresponding to critical mode $(P_C<=P_C^*)$. In this simulation, the performance of each system operating in the design conditions is used as reference. It is noted that the system ERS1 operating with R142b has better COP values than the system ERS3 with R600a and the COP evolution of each system follows that of its ejector entrainment ratio for a fixed geometry (Fig. I.5(a-b)). Indeed, for a fixed value of t_{eE} and an operating ejector in critical mode, the system COP increases when t_{eB} decreases compared to its design value. In addition, a light increase in the system COP is noted for a condenser pressure value $P_C<P_C^*$ compared to its value at the critical point $(P_C=P_C^*)$ which is due to the increase in the refrigerant enthalpy variation between the inlet and the exit of the evaporator.

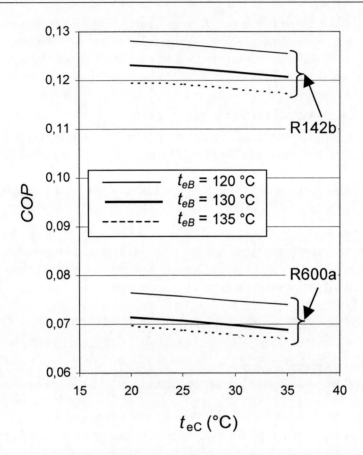

Figure II.12.*COP* of ejector refrigeration systems for air conditioning operating with R142b (ERS1) and R600a (ERS3) and a cold source temperature t_{eE}=10°C [39]

In order to underline the importance of the irreversibilities in the three heat exchangers, the Carnot *COP* is calculated once according to the source temperatures t_{eC}, t_{eB}, t_{eE} and another time according to the corresponding phase change temperatures in the three heat exchangers T_C, T_B, T_E (Fig. II.13).

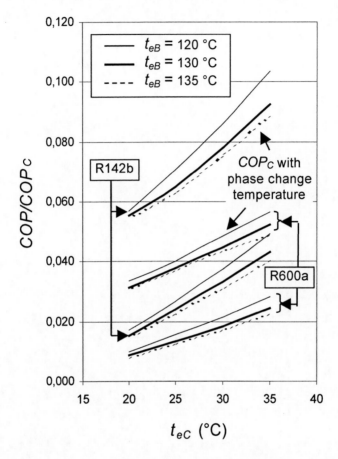

Figure II.13. Effective and Carnot *COP* ratio evolutions of ejector refrigerating systems for air conditioning operating with R142b (ERS1) and R600a (ERS3) and a cold source temperature t_{eE}=10°C [39]

Chapter III

PERFORMANCE IMPROVMENT OF AN EJECTOR REFRIGERATION SYSTEM AND COUPLING

As noted previously, the major drawback of the ejector refrigeration system is its low coefficient of performance mainly due to the low ejector entrainment ratio. Therefore, solutions have been proposed by several researchers, including those relating to the introduction of heat regenerator or mechanical compressor in the basic system cycle.

III.1. HEAT REGENERATOR

In many studies [3,20-21,45-46], two regenerators are added to the conventional system (Fig. I.11) in order to improve the refrigeration efficiency.

The refrigerant temperature is slightly increased in the pre-heater and decreased in the sub-cooler before entering the boiler and the evaporator, respectively. Consequently, the required heat input and the cooling load of the system are reduced. In the study of Dorantès and Lallemand [45], it is shown that the use of heat recovery systems can improve the system *COP* by about 14% with R141b, 22% with R123 and 32% with RC318 (Fig. III.1).

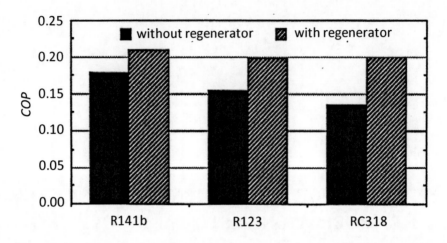

Figure III.1. Cooling efficiency of the jet refrigeration system with and without recovery (T_{Bes}=90°C; T_{Ces}=25°C; T_{Ees}=20°C) [45]

III.2. ADDITION OF S MECHANICAL COMPRESSOR

Performance of an ejector refrigeration system is defined by the entrainment ratio and the critical condenser pressure of the ejector. The only way to increase the secondary flow simultaneously with the ejector back pressure is to increase the evaporator pressure. It is not easy to raise the evaporation temperature because the levels achieved in the conventional ejector refrigeration systems are already not too high. Therefore, to increase the secondary flow pressure without disturbing the refrigeration temperature, two new configurations of efficient use of mechanical power have been proposed by several researchers.

III.2.1. The Booster Assisted Ejector System

The cycle of the booster assisted ejector system [7,11, 19, 54] is not very different from that of the conventional system (Fig. III.2). A low-pressure ratio mechanical compressor is placed between the evaporator outlet and the ejector suction line. Therefore, the suction pressure of the ejector is increased, thus increasing its performance.

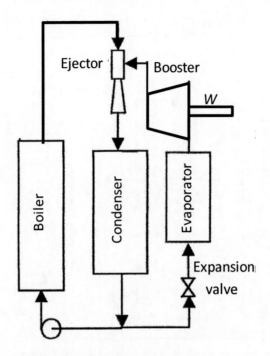

Figure III.2. Booster assisted ejector refrigeration cycle

III.2.2. The Hybrid Compressor-Ejector Refrigeration System

The hybrid compressor-ejector refrigeration system [7,19-21,55-56] is composed of a conventional compression and ejector sub-cycles connected by an intermediate heat exchanger (Fig. III.3). If a single refrigerant is used in the two sub-cycles, the heat exchanger is replaced by a mixing chamber. According to this configuration, pressure ratios across the ejector and the compressor are maintained at a low level. The thermal analysis of hybrid compressor-ejector refrigeration systems shows that the heat load is transferred from the evaporator to the intermediate heat exchanger through the mechanical compressor and then compressed further in the ejector before being rejected to the surrounding at the condenser.

However, it should be noted that the power required by the booster (mechanical compressor) is much higher than that required by the circulation pump and cannot be neglected when evaluating the system performance.

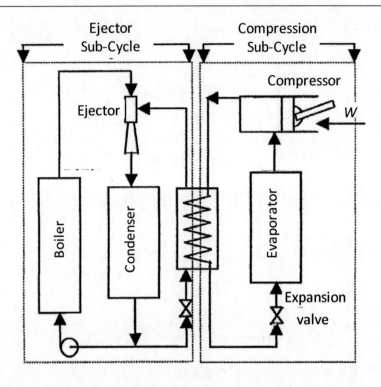

Figure III.3. Hybrid Compression-Jet Refrigeration Cycle [7]

III.3. PERFORMANCE IMPROVMENT OF A CONVENTIONAL VAPOR COMPRESSION REFRIGERATION SYSTEM BY USING AN EJECTOR

Usually, a household refrigerator is equipped with two evaporators (fresh food and freezer compartments) and only one compressor. Tomasek and Radermacher [57] propose to reduce the pressure difference between the freezer evaporator and the condenser by using an ejector (Fig. III.4). Thus, the required specific work to compress the refrigerant is lower and the system *COP* is improved. The performance of a domestic refrigeration system with an ejector has been compared to those of single compressor and two-compressor systems. The results show that the incorporation of an ejector improves by almost 12.4% the *COP* of the ejector-combined system compared to that of a single compressor system, but the two-compressor system has the highest *COP* improvement.

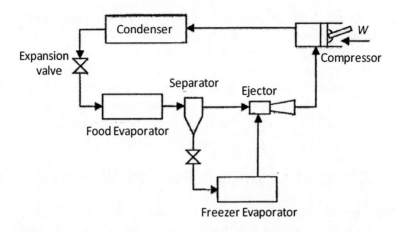

Figure III.4. Schematic diagram of a combined ejector-compression cycle [57]

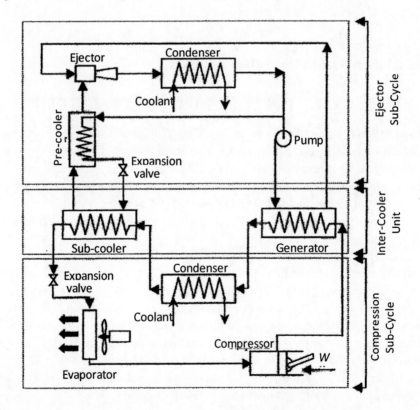

Figure III.5. Schematic diagram of a combined ejector-compression cycle [58]

The *COP* of the conventional vapor compression system can also be improved by increasing the sub-cooling before entering the expansion device. The combined ejector-compression cycle proposed by Huang *et al.* [58] (Fig. III.5) allows using the wasted heat from the superheated vapor in the compression sub-cycle to drive the ejector sub-cycle. The cooling produced by the ejector refrigeration sub-cycle is used to cool the condensate of the compression sub-cycle until a sub-cooled state in order to increase the system *COP*.

III.4. COUPLING OF THE EJECTOR WITH VARIOUS SYSTEMS

The ejector has been the subject of several associations including the association with a solar collector in order to use solar energy as hot source in an ejector refrigeration system or that with an absorption machine in order to increase its coefficient of performance.

III.4.1. Solar Jet Refrigeration System

One of the most important characteristics of an ejector system is its compatibility with the use of a thermal energy source at a low or average temperature. The free energy captured by a solar collector is sufficient to generate the motive fluid in the boiler, which is necessary for its operation. However, according to the temperature levels of the cold source, performance achieved with this system and the cost of initial investment, this system is more suitable for the application of air conditioning system rather than of the refrigeration. In addition, the season of greater cooling need coincides with that of the availability of solar energy.

A solar jet refrigeration system is composed of two sub-systems, which are an ordinary ejector refrigeration system and a solar system. In the basic solar jet refrigeration system (Fig. III.6(a)), the solar cycle acts as the boiler of the refrigeration system. The refrigerant is forced and passes through the absorber of solar collector. This system configuration is abandoned because leaks will cause damage and the boiler pressure controlling becomes difficult in this case. In practice, the solar and ejector refrigeration systems are separated (Fig. III.6(b)). Heat from solar collector is carried out by an intermediate medium and then transferred to the refrigerant by the boiler heat exchanger.

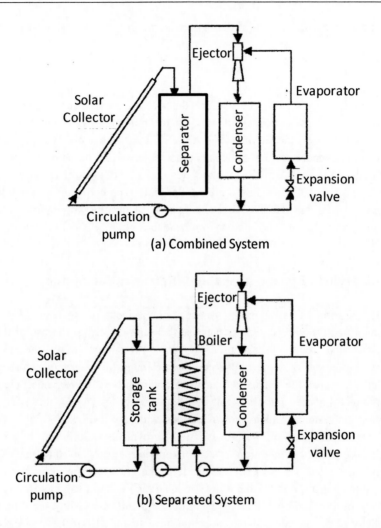

Figure III.6. Schematic diagram of solar jet refrigeration system – (a) Combined system; (b) Separated system [7]

The heat transferring medium should have a boiling point greater than the highest temperature, which can be achieved in the system, a good transfer property and a low viscosity. In order to prevent the intermittence of the solar energy , a backup heater, driven by gas or oil, may be added to the storage tank or the exit of the solar collector to ensure the boiler temperature stability [36,59].

A 7 kW prototype of a solar jet refrigeration system has been installed and tested in an office building in United Kingdom [4].

The overall efficiency of a solar jet refrigeration system is given by [36,59]:

$$COP_{\text{overall}} = COP_{\text{ejector refrigeration system}} \times \eta_{\text{solar system}} \qquad \text{(III.1)}$$

According to the above relation, the overall performance of a solar jet refrigeration system depends not only on the performance of the refrigeration system but also on the thermal efficiency of solar collector. The solar system efficiency depends on the collector type, solar radiation and the operating conditions of the system. For a boiler temperature of the refrigeration system between 80 and 100°C, a single glazed flat-plate type collector with a selective surface is sufficient [36,38]. Parabolic solar concentrating collectors [3,60] and vacuum tube collectors [4,61] are recommended for higher boiler temperature.

III.4.2. Hybrid Ejector-Absorption Refrigeration System

The absorption refrigeration systems, in particular those using LiBr-H_2O, became attractive in recent years because they are adapted to the use of thermal energy wasted from industrial processes or a free energy source such as solar energy. Moreover, they are undoubtedly conform to all environment preserving regulations, since the refrigerant is a natural fluid. However, coefficients of performance for absorption systems are significantly lower than those for vapor compression systems. This has restricted their wide application.

There are several approaches used to improve the performance of a single-effect absorption refrigeration system. Applying ejector to the conventional absorption system is a good solution. Indeed, the incorporation of an ejector to a conventional absorption system allows its COP achieving a value close to that of the COP of a typical double-effect absorption system. Moreover, according to the simplicity of the ejector system, the capital investment cost of the hybrid ejector-absorption refrigeration system is low when compared to other conventional high performance absorption cycle systems.

Various configurations of a hybrid ejector-absorption refrigeration cycle have been proposed by several researchers to increase the performance of the conventional absorption refrigeration system. They aim to increase the absorber pressure or the generator temperature.

In order to increase the absorber pressure at a higher level than the evaporator and thus to reduce the solution concentration, Kuhlenschmidt [62] proposes an absorption system utilizing a two-stage generator similar to that used in a double effect absorption system (Fig. III.7). The low pressure refrigerant from the second-effect generator is used as a motive fluid of the ejector which entrains the refrigerant vapor from the evaporator. Thus, the solution in the absorber is protected against the phenomenon of crystallization, which is harmful to absorption system operation. The major drawback of this system is that a significant part of the refrigerant vapor is directly discharged into the absorber without producing any cooling effect. Therefore, this should affect the system COP. According to our knowledge, there is no value of this system COP available in the literature.

Other researchers [63-64] have used the same configuration of the hybrid ejector-absorption refrigeration system as above with the difference that the strong solution at high pressure returned from the generator is used as the motive fluid of the ejector (Fig. III.8). The experimental tests carried out on this system with the working fluids DMETEG/R21 and DMETEG/R22 have shown a pressure ratio between the absorber and the evaporator of about 1.2.

In order to raise the generator temperature, a new configuration of a hybrid ejector-absorption refrigeration cycle has been proposed [65-66]. An ejector is placed between the generator and the condenser (Fig. III.9). By using a high-temperature heat source, the generator temperature may be increased and the concentration of the strong solution can be kept constant. The generator is thus protected against the phenomenon of crystallization. The ejector uses high-pressure vapor from the generator as the motive fluid. An ejector integrated in this way into the absorption refrigeration system increases the refrigerant flow rate from the evaporator and therefore raises the system cooling capacity. Thus, the COP is higher than that for a conventional system. A prototype of this system has been developed and tested at the University of Sheffield [65]. COP values equal to 0.86-1.04 are achieved. The major drawback of this system is that the generator must operate at very high temperatures (180-210°C) and therefore the equipment corrosion becomes problematic.

In order to eliminate the problem of corrosion, a combined stream jet heat pump-absorption refrigeration cycle has been proposed [67-68]. The steam heat pump (Fig. III.10) allows recovering heat rejected during the refrigerant condensation . The recovered heat is then supplied back to the generator of the absorption refrigeration system. In this case, the solution temperature does not exceed 80°C. For more details on the absorption refrigeration systems refer to [69].

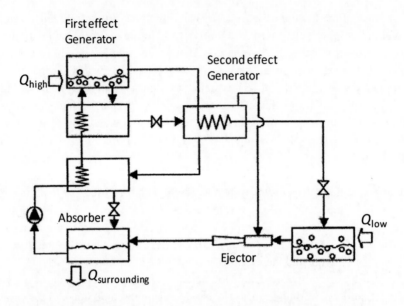

Figure III.7. Schematic diagram of the hybrid ejector-absorption cycle [62]

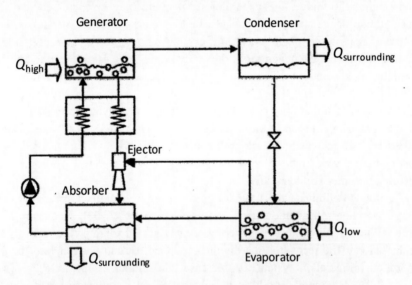

Figure III.8. Schematic diagram of the hybrid ejector-absorption cycle [63-64]

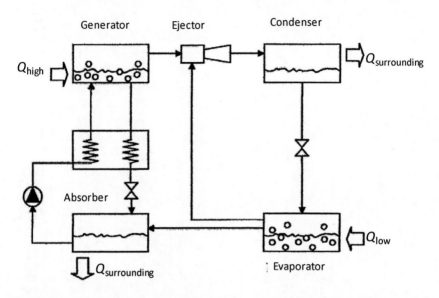

Figure III.9. Schematic diagram of the hybrid ejector-absorption cycle [65-66]

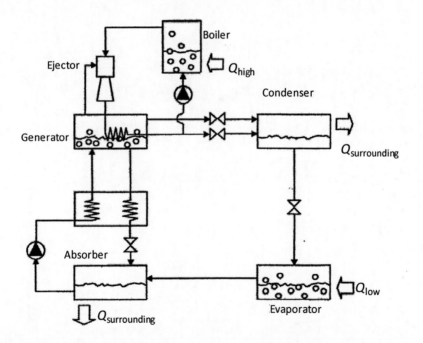

Figure III.10. Schematic diagram of the hybrid ejector-absorption cycle [67-68]

Chapter IV

ENERGETIC EFFICIENCY IMPROVMENT OF THE TRANSCRITICAL CO2 REFRIGERATION SYSTEM BY USING AN EJECTOR AS AN EXPANSION DEVICE

IV.1. INTRODUCTION

In the framework of sustainable development, there has been growing interest in technologies based on ecologically safe 'natural' refrigerants such as air, ammonia, carbon dioxide, hydrocarbons, water, etc. Among these natural refrigerants, carbon dioxide (R744) is considered as an alternative to CFCs, HCFCs and HFCs, particularly in automotive air conditioning and heat pumps, because it is environmentally safe with nearly negligible global warming potential (GWP) and zero ozone depletion potential (ODP). Besides, CO_2 has many excellent advantages in engineering applications, such as non-toxicity, non-inflammability, high volumetric capacity and better heat transfer properties. However, a transcritical CO_2 cycle shows intrinsic disadvantages in air conditioning due to large expansion losses and higher irreversibilities during the gas cooling process. Many researchers have analyzed the performance of the transcritical CO_2 refrigeration cycle in order to identify opportunities to improve the system energy efficiency. By performing a second law analysis, it has been found [70] that the isenthalpic expansion in this cycle is a major contributor to the cycle irreversibilities due to the fact that the expansion process evolves from the supercritical region into the two-phase region. By using vapor compression and transcritical cycle models, the performances of CO_2 and R22 have been compared for residential air-conditioning applications [71]. The results show that the R22 system has a significantly better COP than the CO_2 system when equivalent heat exchangers are used in both cases. An entropy generation analysis shows that the major part of irreversibilities occurs in the expansion valve and in the gas cooler. These irreversibilities are greatly responsible for the low COP of the CO_2 system. Therefore, reduction of losses is become a privileged search track to increase the efficiency of the transcritical CO_2 refrigeration cycle. Thus, a free piston expander-compressor unit has been proposed to recover the expansion process losses [72]. However, implementation of the concept requires a two-stage refrigeration cycle, which complicates flow control devices. After a thermodynamic analysis of various expansion devices for a transcritical CO_2 refrigeration system, a vortex tube expansion device and an expansion work output device have been proposed to recover the expansion losses [73]. By assuming an ideal expansion process, the maximum increase in COP is about 37% in both cases compared to the one using an isenthalpic expansion process. The COP reduces to about 20% when the efficiency for the expansion work output device is 0.5. To achieve the same improvement in COP using a vortex tube expansion device, the efficiency of the vortex tube must be above 0.38. A one-dimensional iterative model for a R12 system with two-phase ejector used instead of a throttling valve to recover expansion work has been presented by Kornhauser [74]. The results show a COP improvement of up to 21% over the conventional cycle with expansion valve. However, improvements in the order of only a few percent have been achieved in the subsequent experiments carried out by Harrell and Kornhauser [75]. Relying on the idea of Kornhauser, Liu *et al.* [76] have performed a thermodynamic analysis of a transcritical R744 system with a two-phase ejector. A theoretical COP improvement between 6 and 14% has been calculated. On the basis of a numerical work similar to that of Liu *et al.*, Jeong *et al.* [77] present a COP improvement up to 22% over the conventional cycle with expansion valve. By including an additional internal heat exchanger in their analysis, Deng *et al.* [10] obtain values of COP improvement of the same magnitude as those presented by Jeong *et al.* More research on R744 two-phase ejector including experimental tests have been carried out by Takeuchi *et al.* [78] and Ozaki *et al.* [79]. Their results show a COP improvement of 20% for an automotive system. Elbel and Hrnjak [80] have obtained experimental results from a transcritical R744 system using an ejector. For the test conditions, the cooling capacity and the COP are improved by up to 8% and 7%, respectively compared to those of a conventional system. Due to difficulties in the ejector throat pressure measurements, a more practical performance metric was introduced by the authors in order to quantify overall ejector efficiencies.

According to this definition, the prototype ejector was able to recover up to 14.5% of the throttling losses. Furthermore, a R744 heat pump with ejector for water heating has been successfully commercialized by a Japanese firm (OEM).

IV.2. STUDY OF AN EJECTOR-EXPANSION TRANSCRITICAL CO2 REFRIGERATION SYSTEM

The objective of this study is to develop a simulation tool to design an ejector in order to incorporate it into a transcritical CO_2 refrigeration system as the principal expansion device and to evaluate the performance improvement. The simulation of CO_2 flow through the ejector, assumed operating in transition mode, is based on the 1-D model of constant-area mixing type. The thermodynamic characteristics of the refrigerant are calculated using the REFPROP® package.

For a cooling capacity to be achieved by the system, section areas of the primary and secondary nozzle throats, the ejector entrainment ratio and the vapor quality at the exit are determined using an assumed value of the pressure lift ratio. Thus the system *COP* and its improvement are compared to that of the basic system. The effect of the gas cooler pressure and the evaporator superheat on the system performance is also investigated.

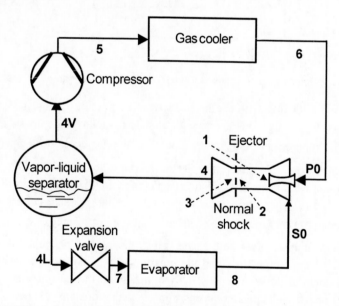

Figure IV.1. Schematic of an ejector expansion refrigeration system

Figs. IV.1 and IV.2 show a schematic of the ejector expansion refrigeration system with a *T-s* diagram illustrating the R744 transcritical cycle. The system includes a compressor, a gas cooler, an ejector, a vapor-liquid separator, an expansion valve and an evaporator.

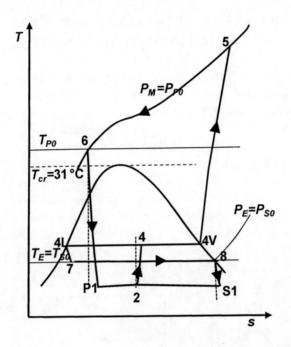

Figure IV.2. Ejector expansion transcritical CO_2 refrigeration cycle in a *T, s* diagram

The sub-critical CO_2 enters the compressor at pressure P_4 at state (4V) and is compressed to the high side pressure P_5 at state (5) with an isentropic efficiency η_{comp}. The supercritical CO_2 is then cooled in the gas cooler to temperature T_6. This motive flow (P) enters the ejector primary nozzle (Fig. II.2), with the thermodynamic characteristics of the stagnation state (T_{P0}, P_{P0}) assumed equal to those of state (6), then undergoes an expansion to the pressure corresponding to that of state (P1), before mixing, with an isentropic efficiency, η_N. At the exit, the primary fluid at supersonic speed drives the secondary fluid (S) from the evaporator at the temperature and the pressure of the stagnation state (T_{S0}, P_{S0}) supposed equal to those of state (8). Then, the driven flow, is accelerated in the secondary nozzle as its pressure drops from P_E to that of state (S1) before mixing, with an isentropic efficiency, η_S. Both fluids (P1) and (S1) mix together in the constant-area section with the mixture final state corresponding to state (2). The pressure increases in the ejector due to a formation of a normal shock in the mixing chamber and to the flow through the diffuser. The diffuser is assumed to have an isentropic efficiency, η_D. The stream leaving the ejector at pressure P_4 in two-phase state (4) flows into the vapor-liquid separator where it is separated into saturated liquid and saturated vapor corresponding to states (4L) and (4V). The saturated liquid enters the expansion valve and expands to pressure P_7 at state (7) before it penetrates in the evaporator to produce the expected refrigerating effect. The saturated vapor (4V) enters the compressor.

IV.3. MATHEMATICAL MODEL

The constant-area mixing model with the same assumptions as those given in § I.1., except that relating to ideal gas behavior which is obviously not valid in this case, is used to analyze the CO_2 flow through the ejector. The isentropic efficiency values for the primary nozzle, the suction chamber and the diffuser are also the same as those given in § I.3.1.

To further simplify the theoretical model of the ejector expansion transcritical CO_2 refrigeration cycle, other assumptions are also made:

- The pressure drops in the gas cooler, the evaporator and the connection tubes are neglected.
- The only heat rejection to the environment from the system is that which takes place in the gas cooler.
- The vapor and liquid streams from the separator are saturated phases.
- The flow through the expansion valve is isenthalpic.
- The compressor isentropic efficiency is equal to 0.75.
- The evaporating temperature is 5°C.
- The refrigerant temperature at the gas cooler outlet is 35°C.

Assuming that the pressure P_1 of the primary and secondary flows (P1) and (S1) before mixing is uniform and determined by the choking condition of the secondary flow at this plane (i.e. $P_1=P_S^*$), the equations for the ejector section before mixing can be identified.

The primary fluid entropy s_{P0} and enthalpy h_{P0} at the ejector inlet are determined from the refrigerant pressure P_5 and temperature T_6 at the gas cooler exit by using Eqs. (II.16-17). The primary stream is accelerated from P_5 to P_P^* corresponding to its choking which occurs at the nozzle throat section. Using the isentropic process, the pressure P_P^* and the section area A_P^* are calculated from Eqs. (II.18-23). Then, the fluid is further accelerated in the primary nozzle divergent corresponding to a pressure drop from P_P^* to P_1. The fluid characteristics at the end of this expansion are determined by using the same procedure. The Mach number M_{P1}, the section area of the nozzle exit A_{P1} and the geometrical parameter φ are calculated from Eqs. (II.1, II.24-26). The expansion process of the secondary fluid in the suction chamber is similar to that of the motive fluid in the primary nozzle convergent. So, for a fixed cooling capacity, the throat cross-section area $A_{S1} = A_S^*$ can be calculated for any value of the primary flow from Eqs. (II.14, II.21, II.27). Therefore, the ejector constant-section area ($A_1 = A_{P1} + A_{S1}$) is also determined.

Mixing of both primary and secondary flows occurs in the cylindrical mixing chamber. The pressure P_2, the velocity V_2 and the enthalpy h_2 are calculated from balance equations of mass (Eqs. (II.30-31) with $A_2 = A_1$), momentum (Eq. (II.6) with $P_{P1} = P_{S1} = P_1 = P_S^*$) and energy (Eq. (II.29)) applied to the mixture between planes 1 and 2. The main geometrical parameter Φ is given by Eq. (II.32). In transition mode, the mixed fluid is compressed by a shock wave which is assumed taking place in the ejector constant section. The characteristics of the mixture downstream (plane 3) are also calculated from the balance equations applied between sections A_2 and A_3 from Eqs. (II.33-36).

Then the mixture flowing in the divergent is again compressed. For a fixed value of the ejector exit pressure P_4 (or the ejector pressure lift ratio r), an isentropic compression (Eqs. (II.37-38)) is supposed to determine the exit state. The corresponding enthalpy h_{4is} is determined from an equation similar to Eq. (II.19) and using Eq. (II.40) the actual enthalpy of the mixture h_4 is found. The output vapor quality can be expressed in two ways, once from Eq. (II.42) and a second time from the equation of mass conservation throughout the facility (Eq. (IV.1)):

$$x_4 = \frac{1}{1+U} \qquad (IV.1)$$

To find the specific work of the compressor w_{comp}, first the isentropic conditions at the compressor outlet are evaluated:

$$s_{4V} = f(P_4, x = 1) \tag{IV.2}$$

The entropy $s_{5,is}$ and the enthalpy $h_{5,is}$ are calculated from Eqs. (II.18-19) using Eq. (IV.2).

From the isentropic efficiency of the compressor η_{comp}, the actual enthalpy at the compressor outlet h_5 can be found:

$$\eta_{comp} = \frac{h_{5,is} - h_{4V}}{h_5 - h_{4V}} \tag{IV.3}$$

with :

$$h_{4V} = f(P_4, x = 1) \tag{IV.4}$$

Then, w_{comp} can be calculated from:

$$w_{comp} = (h_5 - h_{4V}) \tag{IV.5}$$

The COP of the ejector-expansion transcritical CO_2 refrigeration system is evaluated by:

$$COP = \frac{\dot{Q}_E}{\dot{m}_P (h_5 - h_{4V})} \tag{IV.6}$$

The entropy $s_{5b,is}$ and the enthalpy $h_{5b,is}$ of the isentropic process at the compressor outlet for the basic system are determined from Eqs.(II.18-19) using the refrigerant entropy $s_{SE,b}$ at the evaporator exit, calculated by its pressure P_E and its temperature $T_E + \Delta T_E$ from Eq.(II.16).

Using the compressor isentropic efficiency, given by an expression similar to Eq.(IV.3), the actual enthalpy at the compressor outlet h_{5b} can be determined. Then, the specific compressor work of the basic system $w_{comp,b}$ is found:

$$w_{comp,b} = (h_{5b} - h_{sE,b})$$ (IV.7)

with $h_{sE,b}$ is the refrigerant enthalpy at the evaporator exit, calculated by its pressure P_E and its temperature $T_E + \Delta T_E$ from Eq.(II.17).

The CO_2 flow rate $\dot{m}_b$, which should circulate in the basic system for the same cooling capacity $\dot{Q}_E$ as the refrigeration system with ejector expansion, is:

$$\dot{m}_b = \frac{\dot{Q}_E}{(h_{sE,b} - h_6)}$$ (IV.8)

with h_6 is the refrigerant enthalpy at the gas cooler exit, calculated by its pressure P_5 and its temperature T_6 from Eq. (II.17).

The coefficient of performance of the basic system COP_b is evaluated by:

$$COP_b = \frac{\dot{Q}_E}{\dot{m}_b(h_{5b} - h_{sE,b})}$$ (IV.9)

The relative performance of the ejector expansion system to the basic system I is defined as:

$$I = \frac{COP}{COP_b}$$ (IV.10)

IV.4. RESULTS AND DISCUSSION

Fig. IV.3 shows the inputs and outputs of the simulation model which is composed of eight blocks calculated successively. The flowchart of the modeling program is detailed in Fig. IV.4.

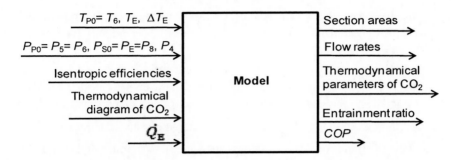

Figure IV.3. Inputs and outputs of the simulation model for an ejector-expansion transcritical CO_2 system

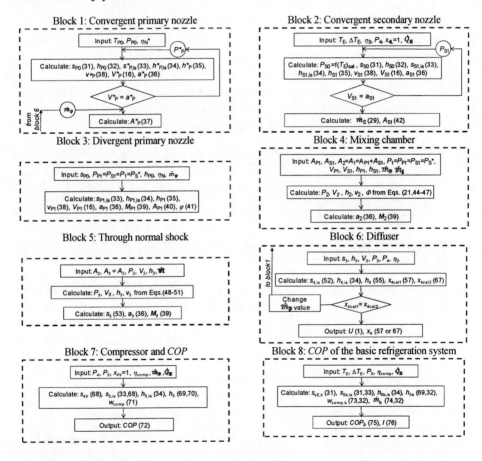

Figure IV.4. The flowchart of the modeling program of an ejector-expansion transcritical CO_2 system

To investigate the characteristics of the transcritical CO_2 system with expansion by ejector, the following standard operating conditions are assumed: $P_5=P_6=P_{P0}=100$ bar, $T_6=35°C$, $T_7=T_8=T_E=5°C$, $\Delta T_E=0°C$.

For various values of the ejector pressure lift ratio (Table IV.1), the Mach numbers at primary nozzle exit and shock wave upstream are constant and equal to 1.64 and 1.29, respectively. The section areas of the primary and secondary nozzles throats A_P^* and A_1 increase with increasing the compression ratio. However, the ratio of these sections Φ decreases significantly, whilst the entrainment ratio U drops as expected. Despite this, for extreme conditions of the cycle, there is a significant improvement (nearly 50%) of the performance coefficient. The comparison of this COP with that of the basic system (constant) highlights the clear benefit of the partial replacement of the expansion valve by the ejector. Indeed, the improvement, equal to 28% for a realistic value of r (equal to 1.2), reaches more than 90% for the maximum considered 1.6.

Table IV.1. Evolution of the ejector geometrical parameters and thermodynamic characteristics of the transcritical CO_2 system with expansion ejector according to the compression ratio r

U	x_4	COP	COP_b	COP/COP_b	A^*_P(m² x10⁶)	A_1(m² x 10⁶)	Φ	M_{p1}	M_2
0.672	0.598	3.707	2.892	1.282	1.179	6.726	5.703	1.641	1.276
0.666	0.600	4.080	2.892	1.411	1.253	7.104	5.669	1.641	1.277
0.656	0.604	4.500	2.892	1.556	1.341	7.540	5.621	1.641	1.279
0.643	0.609	4.975	2.892	1.720	1.451	8.062	5.556	1.641	1.283
0.624	0.616	5.507	2.892	1.904	1.597	8.723	5.462	1.641	1.287

For realistic values of r equal to 1.2 and 1.3, and considering null ΔT_E, the influence of the compressor discharge pressure P_{P0} on:

- the entrainment ratio U and the vapor quality x_4 of the mixture leaving the ejector (Fig. IV.5),
- the COP of a system with ejector-expansion, the basic system COP_b as well as the improvement produced by the ejector COP/COP_b (Fig. IV.6),
- the section area of the primary nozzle throat (Fig. IV.7),
- the geometrical parameter Φ (Fig. IV.8),

is investigated. It is noted the strong decrease of A_p^* and that of A_1, which is less strong. Therefore Φ grows significantly. Unlike U, as would be expected, the performance coefficients of both systems decrease substantially when the pressure P_{P0} increases.

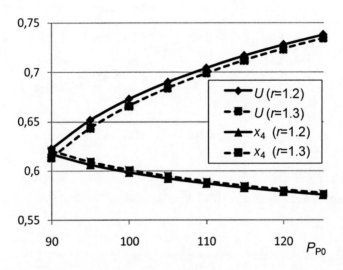

Figure IV.5. Influence of the compressor discharge P_{P0} pressure on the entrainment ratio U and the vapor quality at the ejector output x_4 (with r=1.2, 1.3 and ΔT_E=0K)

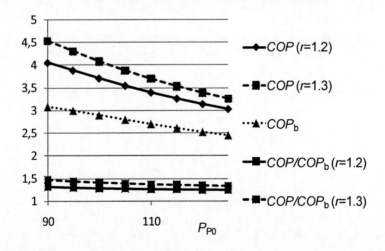

Figure IV.6. Influence of the compressor discharge P_{P0} pressure on the COP, COP_b and COP/COP_b (with r=1.2, 1.3 and ΔT_E=0K)

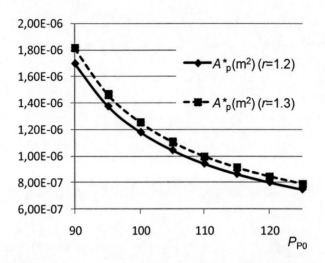

Figure IV.7. Influence of the compressor discharge P_{P0} pressure on the section area of the primary nozzle throat A_P^* (with r=1.2, 1.3 and ΔT_E=0K)

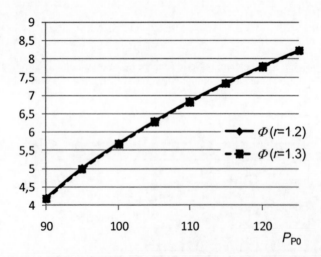

Figure IV.8. Influence of the compressor discharge P_{P0} pressure on the ejector geometrical parameter Φ (with r=1.2, 1.3 and ΔT_E=0K)

For high-side pressure equal to 100 bar, r equal to 1.2 and 1.3, the influence of the evaporator superheat ΔT_E on the system characteristics is also studied. A slight increase of U (and hence a slight decrease of x_4) is noted (Fig. IV.9) with ΔT_E increasing. However, the COP of the ejector-expansion system and that of the basic system COP_b as well as the COP/COP_b ratio decrease slightly when ΔT_E increases (Fig. IV.10).

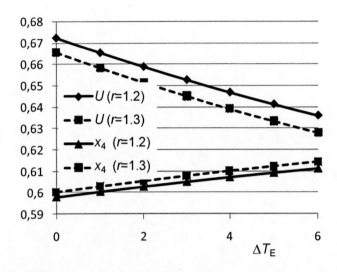

Figure IV.9. Influence of the evaporator superheat ΔT_E on the entrainment ratio U and the vapor quality at the ejector output x_4 (with $r=1.2$, 1.3 and $P_{P0}=100$ bar)

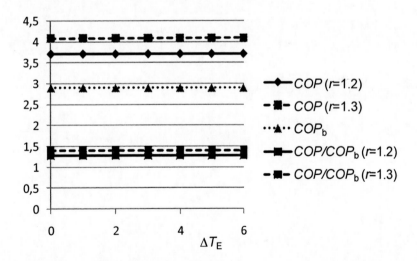

Figure IV.10. Influence of the evaporator superheat ΔT_E on the COP, COP_b and COP/COP_b (with $r=1.2$, 1.3 and $P_{P0}=100$ bar)

CONCLUSION

This work aims to highlight the role of an ejector in energy savings, more particularly in refrigeration systems. To illustrate the application of the ejector in refrigeration two systems are chosen:

- The ejector refrigeration systems which are heat-operated systems where the ejector plays the same role as the compressor unit in a conventional refrigeration system.
- A transcritical CO_2 refrigeration system where the ejector is used as the main expansion device for minimizing the energy losses.

The literature review shows that the ejector operating is not well known. Indeed, if the ejector behavior analysis is often based on the theory of Keenan *et al.*, many mathematical models using new assumptions on mixing and flowing characteristics are always developed. However, the simulation results of only a few of them are experimentally verified.

The ejector is the critical component of a jet refrigeration system and its performance is expressed by two parameters: the entrainment ratio and the pressure lift ratio. The ejector performance depends not only on the system operating conditions, but on its geometry and the nature of the working fluid.

For ejector designing in transition mode, a dynamic correlation of the entrainment ratio as function of the pressure lift and the driving pressure ratios has been developed from experimental tests carried out on CFCs.

By using the same correlation, the influence of the nature of the working fluid (pure or binary mixture) on the system performance has been investigated. The results show that R141b, R123, RC318 and R142b lead to the best system *COP* and exergetic efficiency; moreover the use of a binary mixture does not necessarily increases the system performance. Indeed, the system *COP* and exergetic efficiency decrease when the mixture is strongly zeotropic (R22/RC318) compared to those of one or other of the two pure fluids and they increase when the mixture is moderately zeotropic (R134a/R142b) or almost azeotropic (R134a/R152a) compared to those of the two pure fluids. Also the use of mixtures having a molecular weight in the power loop different from that of the refrigerant loop, owing to a distillation at the ejector output, has been studied.

Using an intermediate fluid (HCFC: R142b) and a natural fluid (isobutane: R600a), a global model of the ejector refrigeration system, based on those of its components, has been developed. For air conditioning applications, the system performance is investigated in design and off-design conditions. The simulation results suggest the following comments:

- At fixed geometry and cold source temperature, the system COP with ejector operating at critical mode ($P_C <= P_C^*$) decreases when the hot source temperature is higher than that of the system design. Therefore, it is better to dimension all the system components at the highest possible temperature of the hot source in order to guarantee better performance in a use at lower temperature. Besides, we can conclude that the system COP variation according to the hot source temperature does not follow that of Carnot $COP,$ which increases with this temperature, leading to a reduction of the system ratio COP/COP_C.

- For ejector refrigeration systems designed for the working fluids R142b and R600a at the same temperatures of the three heat sources, the system COP operating with R142b is better. This is due to the R142b properties , in particular its molecular weight which is almost twice more important than that of R600a.

One of the significant characteristics of a jet cooling system is that the required hot source temperature is compatible with the utilization of low grade energy such as solar energy to drive the system.

Even though the principal drawback of the ejector refrigeration system, which impedes to its broad diffusion, is its low COP, the COP improvement by combining ejector to other types of refrigeration systems (vapor compression or absorption systems) is interesting.

For a cooling capacity of 10 kW and typical operating conditions of air conditioning, an ejector has been designed in transition mode for a fixed value of the pressure lift ratio, using a constant-area mixing model. Then a thermodynamic cycle analysis of the transcritical CO_2 system with expansion by ejector is performed using the same model for the ejector. For gas cooler pressure fixed at 100 bar and realistic pressure lift ratio equal to 1.2 and 1.3, the expansion by ejector improves the COP by more than 28 to 41%, respectively, compared to the basic cycle. Moreover, for a fixed lift pressure ratio r, the ejector entrainment ratio increases while the coefficient of performance of the system with ejector-expansion and that of the basic system decrease significantly when the pressure of the gas cooler increases. The simulation results also show that the profit achieved by the introduction of the ejector in the basic system decreases. Indeed, for r equal to 1.2, the COP improving is 31.6% at 90 bar and is less than 24% at 125 bar. However, the evaporator superheat has little influence on system performance.

REFERENCES

[1]   Chen, L.T. A new ejector-absorber cycle to improve the COP of an absorption system. *Appl Energy*, 2001, *vol* 30, 37-41.

[2]   Srikhirin, P.; Aphornratana,S. ; Chungpaibulpatana, S. A review of absorption refrigeration technologies. *Renew Sustain Energy Rev*, 2001, *vol* 5, 343-372.

[3]   Wolpert, J.L.; Riffat, S.B. 13 kW cooling hybrid solar/gas ejector air conditioning system. In: *First International Conference on Sustainable Energy Technologies*, 2002.

[4]   Nguyen, V.M.; Riffat, S.B.; Doherty, P.S. Development of a solar-powered passive ejector cooling system. *Appl Thermal Eng*, 2001, *vol* 21, 157-168.

[5]   Lu, L.T. *Etudes théorique et expérimentale de la production de froid par machine tritherme à éjecteur de fluide frigorigène*. PhD Thesis, Institut National Polytechnique, Grenoble, FR, 1986.

[6]   Nadhi, E. *Etude paramétrique expérimentale des caractéristiques du système tritherme à éjecteur*. PhD Thesis, Institut National des Sciences Appliquées, Lyon, FR, 1989.

[7]   Chunnanond, K.; Aphornratana, S. Ejectors : applications in refrigeration technology. *Renew Sustain Energy Rev*, 2004, *vol* 8, 129-155.

[8]   Li, D.; Groll, E.A. Transcritical CO_2 refrigeration cycle with ejector-expansion device. *Int J Refrig*, 2005, *vol* 28, 766-773.

[9]   Liu, F.; E.A. Groll, E.A. Recovery of throttling losses by a two-phase ejector in a vapor compression cycle. *ARTI- 10110-01Report*, 2008.

[10]   Deng, J.; Jiang, P.; Lu, T.; Lu, W. Particular characteristics of transcritical CO_2 refrigeration cycle with an ejector. *Appl Thermal Eng*, 2007, *vol* 27, 381-388.

[11]   Stoecker, W.F. *Steam-jet refrigeration*; Mc Graw-Hill: Boston, MA, US, 1958.

[12] Rogdakis, E.D.; Alexis, G.K. Design and parametric investigation of an ejector in an air-conditioning system, *Appl Thermal Eng*, 2000, *vol* 20, 213-226.

[13] Selvaraju, A.; Mani, A. Analysis of a vapor ejector refrigeration system with environment friendly refrigerants. *Int J Therm Sciences*, 2004, *vol* 43, 915-921.

[14] Keenan, J.H.; Neumann, E.P.; Lustwerk, F. An investigation of ejector design by analysis and experiment. *ASME J Appl Mech Trans*, 1950, *vol* 72, 299-309.

[15] Munday, J.T.; Bagster, D.F. A new theory applied to steam jet refrigeration. *Ind Eng Chem Process Res and Dev*, 1977, *vol* 16 (4), 442-449.

[16] Keenan, J.H.; Neumann, E.P. A simple air ejector. *ASME J Appl Mech Trans*, 1942, *vol* 64, 75-81.

[17] Yapici, R.; Ersoy, H.K. Performance characteristics of the ejector refrigeration system based on the constant area ejector flow model. *Energy Conv Mngmnt*, 2005, *vol* 46, 3117-3135.

[18] Rogdakis, E.D.; Alexis, G.K. Investigation of ejector design at optimum operating condition. *Energy Conv Mngmnt*, 2000, *vol* 41, 1841-1849.

[19] Sokolov, M.; Hershgal, D. Enhanced ejector refrigeration cycles powered by low grade heat. Part 1. Systems characterization. *Int J Refrig*, 1990, *vol* 13, 351-356.

[20] Sokolov, M.; Hershgal, D. Enhanced ejector refrigeration cycles powered by low grade heat. Part 2. Design procedures. *Int J Refrig*, 1990, *vol* 13, 357-363.

[21] Sokolov, M.; Hershgal, D. Enhanced ejector refrigeration cycles powered by low grade heat. Part 3. Experimental results. *Int J Refrig*, 1990, *vol* 14, 24-31.

[22] Sun, D.W. Variable geometry ejectors and their applications in ejector refrigeration systems. *Energy*, 1996, *vol* 21 (10), 919-929.

[23] Eames, I.W.; Aphornratana, S.; Haider, H. A theoretical and experimental study of a small-scale steam jet refrigerator. *Int J Refrig*, 1995, *vol* 18 (6), 378-386.

[24] Aly, N.H.; Karameldin, A.; Shamloul, M.M. Modeling and simulation of steam jet ejectors. *Desalination*, 1999, *vol* 123, 1-8.

[25] Al-Khalidy, N.; Zayonia, A. Design and experimental investigation of an ejector in air-conditioning and refrigeration system. *ASHRAE Trans*, 1995, *vol* 101(2), 383-391.

[26] Huang, B.J.; Chang, J.M.; Wang, C.P.; Petrenko, V.A. A 1-D analysis of ejector performance. *Int. J. Refrig*, 1999, *vol* 22, 354-364.

[27] Cizingu, K.; Mani, A.; Groll, M. Performance comparison of vapor jet refrigeration system with environment friendly working fluids. *Appl Thermal Eng*, 2001, *vol* 21, 585-598.

[28] El-Dessouky, H.; Ettouney, H.; Alatiqi, I.; Al-Nuwaibit, G. Evaluation of steam jet ejector. *Chem Engng Process*, 2002, *vol* 41, 551-561.

[29] Chunnanond, K.; Aphornratana, S. An experimental investigation of steam-ejector refrigerator, the analysis of pressure profile along ejector. In: *The asia-Pacific conference on sustainable Energy and Environment Technologies*, 2003, pp 184-188.

[30] Chen, Y.M.; Sun, C.Y. Experimental study of the performance characteristics of a steam ejector refrigeration system. *Exp Fluid Sci*, 1997, *vol* 15, 384-394.

[31] Riffat, S.B.; Gan, G.; Smith, S. Computational fluid dynamics applied to ejector heat pump. *Appl Thermal Eng*, 1996, *vol* 16 (4), 291-297.

[32] Rusly, E.; Aye, Lu.; Charters, W.W.S.; Ooi, A.; Pianthong, K. Ejector CFD modeling with real gas model. In: *Mechanical Engineering Network of Thailand, the 16th Conference*, 2002, pp 150-155.

[33] Desevaux, P. A method for visualizing the mixing zone between co-axial flows in an ejector. *Optics Lasers Engng*, 2001, *vol* 35, 317-323.

[34] Huang, B.J.; Jiang, C.B.; Hu, F.L. Ejector performance characteristics and design analysis of jet refrigeration system. *Trans ASME*, 1985, *vol* 107, 792-802.

[35] Riffat, S.B.; Holt, A. A novel heat pipe/ejector cooler. *Appl Thermal Eng*, 1998, *vol* 18 (3-4), 93-101.

[36] Khattab, N.M.; Barakat, M.H. Modeling the design and performance characteristics of solar steam-jet cooling for comfort air conditioning. *Solar Energy*, 2002, *vol* 73 (4), 257-267.

[37] Srisastra, P.; Aphornratana, S. A circulating system for a steam jet refrigeration system. *Appl Thermal Eng*, 2005, *vol* 25, 2247-2257.

[38] Huang, B.J.; Chang, J.M. Empirical correlation of ejector design. *Int J Refrig*, 1999, *vol* 22, 379-388.

[39] Boumaraf, L.; Lallemand, A. Modeling of an ejector refrigerating system operating in dimensioning and off-dimensioning conditions with the working fluids R142b and R600a. *Appl Thermal Eng*, 2009, *vol* 29, 265-274.

[40] Dorantès, R.J. *Performances théoriques et expérimentales d'une machine frigorifique tritherme à éjecto-compression. Influence de la nature du fluide de travail. Analyses énergétique et exergétique.* PhD Thesis, Institut National des Sciences Appliquées, Lyon, FR, 1992.

[41] Eames, I.W.; Wu, S.; Worall, M.; Aphornratana, S. An experimental investigation of steam ejectors for application in jet-pump refrigerators powered by low-grade heat. *Proc. Inst Mech Eng A*, 1999, *vol* 213, 351-361.

[42] Aphornratana, S.; Eames, I.W. A small capacity steam-ejector refrigerator: Experimental investigation of a system using ejector with movable primary nozzle. *Int J Refrig*, 1997, *vol* 20 (5), 352-358.

[43] Work, L.T.; Miller, A. Factor C in the performance of ejectors, as a function of molecular weights of vapors. *Ind Eng Chem*, 1940, *vol* 32 (9), 1241-1243.

[44] Work, L.T.; V. Haedrich, V. Performances of ejectors as a function of the molecular weights of vapors. *Ind Eng Chem*, 1939, *vol* 31 (4), 464-477.

[45] Dorantès, R.J.; A. Lallemand, A. Prediction of performance of a jet cooling system operating with pure refrigerants or non-azeotropic mixtures. *Int J Refrig*, 1995, *vol* 18 (1), 21-30.

[46] Boumaraf, L.; Lallemand, A. Performance analysis of a jet cooling system using refrigerant mixtures. *Int J Refrig*, 1999, *vol* 22, 580-589.

[47] Defrates, L.A.; Hoerl, A.E. Optimum design of ejectors using digital computers. *Chem Eng Prog Symp Ser*, 1959, *vol* 55 (21), 43-51.

[48] Dorantès, R.J.; Moszkowicz, P.; Lallemand, A. Use of non-azeotropic refrigerant mixtures in an ejector air conditioning system. In: *CLIMA 2000 Conf London*, 1993; Vol. 311, pp 1-9.

[49] [49] Ozisik, M.N. *Basic heat transfer*; Mac Graw-Hill Book Company: New York, US, 1977, 571 p.

[50] Nusselt, M. Die Oberflächen Kondensation der Wasserdämfes. *Zeitschrift V D I*, 1916, *vol* 60, 541-546.

[51] Chen, J.C. A correlation for boiling heat transfer to saturated fluids in convective flow. *I Ec Process Design Develop*, 1966, *vol* 5 (3), 322-329.

[52] Rigot, G. Prédiction de la charge en fluide frigorigène des échangeurs fonctionnant en régime diphasique et établi. In: *19th Int Congress Refrigeration*, Den Hague, NL, 1995.

[53] Boumaraf, L.; Lallemand, A. Dimensionnement d'une machine de climatisation tritherme dans les conditions de fonctionnement optimales de son éjecteur utilisant le R142b et R600a. In: *COFRET'06*, Timisoara, RO, 2006.

[54] Dorantès, R.J.; Estrada, C.A.; Pilatowsky, I. Mathematical simulation of a solar ejector-compression refrigeration system. *Appl Thermal Eng*, 1996, *vol* 16 (8-9), 669-675.

[55] Hernandez, J.I.; Dorantès, R.J.; Best, R.; Estrada, C.A. The behavior of a hybrid compressor and ejector refrigeration system with refrigerants 134a and 142b. *Appl Thermal Eng*, 2004, *vol* 24, 1765-1783.

[56] Goktun, S. Optimization of irreversible solar assisted ejector-vapor compression cascaded system. *Energy Conver Mngmnt*, 2000, *vol* 41, 625-631.

[57] Tomasek, M.L.; Radermacher, R. Analysis of a domestic refrigeration cycle with an ejector. *ASHRAE Trans*, *vol* 101 (45), 1431-1438.

[58] Huang, B.J.; Petrenko, V.A.; Chang, J.M.; Lin, C.P.; Hu, S.S. A combined-cycle refrigeration system using ejector cooling cycle as the bottom cycle. *Int J Refrig*, 2001, *vol* 24, 391-399.

[59] Huang, B.J.; Chang, J.M.; Petrenko, V.A.; Zuck, K.B. A solar ejector cooling using refrigerant R141b. *Solar Energy*, 1998, *vol* 64, 223-226.

[60] Al-Khalidy, N. Experimental investigation of solar concentrators in a refrigerant ejector refrigeration machine. *Int J Energy Res*, 1997, *vol* 21, 1123-1131.

[61] Huang, B.J.; Petrenko, V.A.; Samofatov, I.Y.; Shchetinina, N.A. Collector selection for solar ejector cooling system. *Solar Energy*, 2001, *vol* 7 (4), 269-274.

[62] Kuhlenschmidt, D. Absorption refrigeration system with multiple generator stages. *US patent no* 3717007.

[63] Chung, H.M.; Huor, H.; Prevost, M.; Bugarel, R. Domestic heating application of an absorption heat pump, directly fired heat pump. In: *Proceedings International Conference University of Bristol*, 1984.

[64] Chen, L.T. A new ejector-absorber cycle to improve the COP of an absorption system. *Appl Energy*, 1998, *vol* 30, 37-41.

[65] Aphornratana, S. *Theoretical and experimental investigation of a combine ejector-absorption refrigerator*. PhD thesis, University of Sheffield, UK, 1994.

[66] Sun, D.W.; Eames, I.W.; Aphornratana, S. Evaluation of a novel combined ejector-absorption refrigeration cycle-I: computer simulation. *Int J Refrig*, 1996, *vol* 19(3), 172-180.

[67] Wu, S.; Eames, I.W. A novel absorption-recompression refrigeration cycle. *Appl Thermal Eng*, 1998, *vol* 19, 1149-1157.

[68] Eames, I.W.; Wu, S. A theoretical study of an innovative ejector powered absorption- recompression cycle refrigerator. *Int J Refrig*, 2000, *vol* 23, 475-484.

[69] Srikhirin, P.; Aphornratana, S.; Chungpaibulpatana, S. A review of absorption refrigeration technologies. *Renew Sustain Energy Rev*, 2001, *vol* 5, 343-372.

[70] Robinson, D.M.; Groll, E.A. Efficiencies of transcritical CO_2 cycles with and without an expansion turbine. *Int J Refrig*, 1998, *vol* 21 (7), 577-589.

[71] Brown, J.S.; Kim, Y.; Domanski, P.A. Evaluation of carbon dioxide cycle carbon dioxide as R22 substitute for residential air-conditioning. *ASHRAE Trans*, 2002, *vol* 108 (Part 2), 954-964.

[72] Heyl, P.; Kraus, W.E.; Quack, H. Expander-compressor for a more efficient use of CO_2 as refrigerant. In: *Natural working fluids, IIR- Gustav Lorentzen conference*, Oslo, NO, 1998, pp 240-248.

[73] Li, D.; Baek, J.S.; Groll, E.A.; Lawless, P.B. Thermodynamic analysis of vortex tube and work output expansion devices for the transcritical carbon dioxide cycle. In: *Fourth IIR-Gustav Lorentzen conference on natural working fluids*, Purdue University, US, 2000, pp 433-440.

[74] [74] Kornhauser, A.A. The use of an ejector as a refrigerant expander. In: *Proceedings of the 1990 USNCR/IIR-Purdue Refrigeration Conference*, West Lafayette, IN, US, 1990, pp. 10-19.

[75] Harrell, G.S.; Kornhauser, A.A. Performance tests of a two-phase ejector. In: *Proceedings of the 30th Intersociety Energy Conversion Engineering Conference*, Orlando, FL, US, 1995, pp 49-53.

[76] Liu, J.P.; Chen, J.P.; Chen, Z.J. Thermodynamic analysis on transcritical R744 vapor-compression/ejection hybrid refrigeration cycle. In: *Preliminary proceedings of the fifth IIR-Gustav Lorentzen conference on natural working fluids*, Guangzhou, CN, 2002, pp 184-188.

[77] Jeong, J.; Saito, K.; Kawai, S.; Yoshikawa, C.; Hattori, K. Efficiency enhancement of vapor compression refrigerator using natural working fluids with two-phase flow ejector. In: *CD-ROM Proceedings of the 6th IIR-Gustav Lorentzen Conference on Natural Working Fluids*, Glasgow, UK, 2004.

[78] Takeuchi, H.; Nishijima, H.; Ikemoto, T. World's first high efficiency refrigeration cycle with two-phase ejector: 'Ejector cycle'. In: *SAE World Congress Exhibition*, Detroit, MI, US, 2004.

[79] Ozaki, Y.; Takeuchi, H.; Hirata, T. Regeneration of expansion energy by ejector in CO_2 cycle. In: *CD-ROM Proceedings of the 6th IIR-Gustav Lorentzen Conference on natural working fluids*, Glasgow, UK, 2004.

[80] Elbel, S.; Hrnjak, P. Experimental validation of a prototype ejector designed to reduce throttling losses encountered in transcritical R744 system operation. *Int J Refrig*, 2008, *vol* 31, 411-422.

INDEX